Integrated Geophysical Methods for Shallow Aquifers Characterization and Modelling

Integrated Geophysical Methods for Shallow Aquifers Characterization and Modelling

Editors

Francisco Javier Alcalá
Maria Catarina Paz
Pedro Martínez-Pagán
Fernando Monteiro Santos

MDPI • Basel • Beijing • Wuhan • Barcelona • Belgrade • Manchester • Tokyo • Cluj • Tianjin

MDPI

Editors

Francisco Javier Alcalá
Spanish National Research Council (EEZA-CSIC)
Spain

Maria Catarina Paz
Polytechnic Institute of Setúbal
Portugal

Pedro Martínez-Pagán
Technical University of Cartagena
Spain

Fernando Monteiro Santos
University of Lisbon
Portugal

Editorial Office
MDPI
St. Alban-Anlage 66
4052 Basel, Switzerland

This is a reprint of articles from the Special Issue published online in the open access journal *Applied Sciences* (ISSN 2076-3417) (available at: https://www.mdpi.com/journal/applsci/special_issues/shallow_aquifers_characterization_modelling).

For citation purposes, cite each article independently as indicated on the article page online and as indicated below:

LastName, A.A.; LastName, B.B.; LastName, C.C. Article Title. *Journal Name* **Year**, *Volume Number*, Page Range.

ISBN 978-3-0365-3385-8 (Hbk)
ISBN 978-3-0365-3386-5 (PDF)

Contents

About the Editors

Francisco Javier Alcalá (Dr.) is a Senior Researcher at the Spanish National Research Council. His research interests focus on human and global interactions underlying problems of groundwater quantity and quality for strategies of diagnosis, operation, mitigation, and protection at different spatiotemporal scales and climate scenarios. He has experience as a Principal Investigator in National and International R&D projects. He is the author of 54 high-impact SCI papers, and has significant activity as a reviewer and editor, and representation in international committees and forums.

Maria Catarina Paz (Dr.) is a Researcher at the Polytechnic Institute of Setúbal in Portugal. Her research interests include the use of near-surface geophysical and computational techniques to evaluate the impacts of global change on ecosystem functions, such as hydrogeological, soil, permafrost, and biological processes. She has professional experience in applied geophysics in the industry. She has authored several SCI papers and is an active peer-reviewer.

Pedro Martínez-Pagán is an Associate Professor at the technical University of Cartagena in Spain. His research interests include applications of near-surface electrical and electromagnetic methods for environmental, mining and hydrogeological studies. He has used seismic methods, such as the MASW, in microzoning studies aimed to assess urban seismic risks. He has authored 24 high-impact SCI papers and is active as a peer-reviewer. He gives lectures in the MSc and PhD programs for Civil and Mining Engineering students.

Fernando Monteiro Santos is an Associate Professor in Applied Geophysics at the University of Lisbon in Portugal. He is currently conducting research and developing activities on the inversion of electromagnetic data. He has participated in numerous research projects and in collaboration with the industry. He is the author of more than 120 papers and is an active peer-reviewer of R&D actions.

Preface to "Integrated Geophysical Methods for Shallow Aquifers Characterization and Modelling"

This book collects recent and original contributions in the field of integrated geophysical methods for the characterization and modeling of shallow aquifers. The readers will find the contributions both interesting and inspiring when exploring the integration of different electrical, electromagnetic, and seismic geophysical techniques, and/or the combination of geophysical techniques and external data (e.g., geotechnical soundings logs, geochemical tracers, and physical parameters) to reduce the ambiguity of interpretations and validate the geophysical models. The findings and methods presented in these original contributions seek to be of interest concerning some of the problems associated with achieving a holistic strategy to define aquifer geometry and some transient groundwater features, which are necessary data to implement numerical tools for the modeling of the dynamics of groundwater quantity (flow) and quality (salinity and pollution). The Editors envision that these contributions will also be of interest to researchers and practitioners and will help to identify further research initiatives.

Francisco Javier Alcalá, Maria Catarina Paz, Pedro Martínez-Pagán, Fernando Monteiro Santos
Editors

applied sciences

MDPI

Editorial

Integrated Geophysical Methods for Shallow Aquifers Characterization and Modelling

Francisco Javier Alcalá [1,*], Maria Catarina Paz [2], Pedro Martínez-Pagán [3] and Fernando Monteiro Santos [4]

1 Experimental Station of Arid Zones, Spanish National Research Council (EEZA-CSIC), 04120 Almería, Spain
2 CIQuiBio, Barreiro School of Technology, Polytechnic Institute of Setúbal, 2839-001 Lavradio, Portugal; catarina.paz@estbarreiro.ips.pt
3 Department of Mining and Civil Engineering, Universidad Politécnica de Cartagena, 30203 Cartagena, Spain; p.martinez@upct.es
4 Instituto Dom Luiz, Faculdade de Ciências, Universidade de Lisboa, 1749-016 Lisboa, Portugal; fasantos@fc.ul.pt
* Correspondence: fjalcala@eeza.csic.es; Tel.: +34-950-281-045

Citation: Alcalá, F.J.; Paz, M.C.; Martínez-Pagán, P.; Monteiro Santos, F. Integrated Geophysical Methods for Shallow Aquifers Characterization and Modelling. *Appl. Sci.* **2022**, *12*, 2271. https://doi.org/10.3390/app12052271

Received: 2 February 2022
Accepted: 17 February 2022
Published: 22 February 2022

1. Introduction

Aquifers stock about 31.4% of the freshwater on the Earth, provide about 50% of current potable water supply, constitute the sole water source in many areas, support groundwater-dependent ecosystems, and present more resilience than surface watercourses to the negative effects of climate change and anthropogenic activities. However, groundwater use is growing, and signs of degradation are increasingly reported.

Groundwater research is increasingly using geophysical prospecting surveys for aquifer characterization in different hydrogeological environments. They have become a part of the holistic strategy to define aquifer geometry and some transient groundwater features, which are necessary data to implement numerical tools for the modelling of groundwater quantity (flow) and quality (salinity, pollution) dynamics. Geophysical prospecting techniques are non-invasive, usually inexpensive to apply, and useful for providing the accurate and fast subsurface information required for detailed groundwater research over multiple observation scales.

Electrical, electromagnetic, and seismic geophysical techniques are widely used for aquifer characterization. The first two are typically used to deduce aquifer geometry and certain transient groundwater features such as piezometric level, freshwater–saltwater interface, and pore water conductivity, whereas the latter are mostly used to deduce aquifer geometry and certain steady aquifer hydraulic parameters. The integrated use of different techniques reduces the ambiguity of interpretations, especially when the conductive structures and pore-filling fluids (natural and human-induced) are subjected to the temporal dynamics of water content and dissolved ions. Integration can also be referred to using external data (e.g., geotechnical soundings logs, geochemical tracers, physical parameters) to improve and/or validate the geophysical models. Different scientific software platforms with friendly interfaces, robust algorithms for data inversion, and tools for uncertainty analysis are available.

In this broad hydro-geophysical framework, this Special Issue aimed to attract specialized researchers in applied geophysical prospecting techniques for groundwater research, with a special focus on near-surface geophysical prospecting applications for shallow groundwater research. The accepted papers included (i) geophysical prospecting surveys as a part of the holistic strategy for aquifer conceptualization and modelling, (ii) integrated near-surface geophysical prospecting techniques and time-lapse approaches to reduce the ambiguity of hydrogeological interpretations, (iii) experimental field operational designs, and (iv) case studies surveying saturated and unsaturated media for methodological and conceptual purposes. Other papers contributed to the state of the art of the geophysical techniques through specific study cases covering (i) hydrogeological environments such as groundwater-dependent ecosystems (GWEs) and urbanized areas in different countries; (ii) aquifer typologies in coastal and inland areas such as weathered Paleozoic crystalline bedrocks, Mesozoic to Neogene carbonated terrains, Neogene and Quaternary volcanic formations, and Neogene and Quaternary detrital sediments; and (iii) climates including humid, sub-humid and semiarid. The used techniques were (i) electrical, such as electrical resistivity tomography (ERT), floating electrical resistivity (FER), and vertical electrical sounding (VES); (ii) electromagnetic such as ground penetrating radar (GPR), time-domain electromagnetic (TDEM), and frequency-domain electromagnetic (FDEM), including low-frequency magnetotelluric sounding (L-MTS); and (iii) seismic, such as multichannel analysis of surface waves (MASW), refraction microtremor (REMI), and vertical seismic refraction (VSR).

2. Contributions

Since the call for papers was announced in June 2019, a total of 11 manuscripts were received. After a rigorous review process, eight papers have been accepted for publication [1–8]. To gain a better insight into the essence of the Special Issue, we offer brief highlights of the published papers below.

The paper "Integrated MASW and ERT Imaging for Geological Definition of an Unconfined Alluvial Aquifer Sustaining a Coastal Groundwater-Dependent Ecosystem in Southwest Portugal" [1] integrates MASW and time-lapse ERT to define aquifer geometry and identify transient groundwater features of the Cascalheira Stream Basin Holocene alluvial aquifer. This aquifer contributes to the Santo André Lagoon, which is part of a coastal GDE in southwest Portugal. This paper contributes a way to disambiguate geological structures in low electrical resistivity (ER) environments, such as coastal areas. The methodology serves to improve the design of shallow groundwater research.

The paper "Geophysical Characterization of Aquifers in Southeast Spain Using ERT, TDEM, and Vertical Seismic Reflection" [2] assesses the effectiveness of different geophysical prospecting techniques to study the Loma de Úbeda Jurassic dolomite-confined thick aquifer in southern Spain. The VSR technique identified the high-amplitude seismic reflectors of the confined structure (aquifer) from the low-amplitude seismic reflectors of the clay-rich confining lower and upper structures. The ERT technique identified lateral changes in facies and small faults. The TDEM technique complemented the VSR and ERT techniques to widen the prospecting depth range.

The paper "Characterization of a Shallow Coastal Aquifer in the Framework of a Subsurface Storage and Soil Aquifer Treatment Project Using Electrical Resistivity Tomography (Port de la Selva, Spain)" [3] couples ERT surveys with implicit modelling tools to identify aquifer geometry and characterize the saltwater intrusion in the Port de la Selva shallow alluvial aquifer in northeast Spain. With the aim to monitor the effects of water percolation through infiltration ponds, the proposed approaches can improve the commitment of stakeholders to the benefits of soil–aquifer treatment procedures for water reuse as an additional non-conventional water source.

The paper "Identifying Changes in Sediment Texture along an Ephemeral Gravel-Bed Stream Using Electrical Resistivity Tomography 2D and 3D" [4] combines the ERT technique with datasets from borehole logs to analyze the inner geometry of channel cross-sections in a gravel-bed ephemeral stream in southeast Spain. The ERT models were correlated with sediment texture data, such as grain size distribution, effective grain size, sorting, and particle shape (Zingg's classification), in order to integrate the horizontal and vertical ER distributions into a 3D model, thus facilitating the identification of layers according to differential sediment supply at the basin scale.

The paper "Combining of MASW and GPR Imaging and Hydrogeological Surveys for the Groundwater Resource Evaluation in a Coastal Urban Area in Southern Spain" [5] conceptualizes and evaluates the groundwater resource in Adra town in southern Spain, a coastal urban area hydrologically influenced by peri-urban irrigation agriculture. The study included a geological, hydrological, and hydrogeological data compilation, and MASW and GPR surveys to define shallow geological structures and some hydrogeological features. The paper also illustrates how urban groundwater reuse can alleviate the pressure on the currently overexploited regional aquifers.

The paper "Temporal and Spatial Groundwater Contamination Assessment Using Geophysical and Hydrochemical Methods: The Industrial Chemical Complex of Estarreja (Portugal) Case Study" [6] presents data from several geophysical and hydrochemical campaigns carried out to monitor groundwater contamination in the industrial chemical complex of Estarreja in northern Portugal over a period of 30 years. With more than a half-century in operation, this complex has left serious environmental liabilities in its influencing area. Findings from geophysical surveys (using the FDEM technique) are part of the research strategy for soil and groundwater remediation.

The paper "Usefulness of Compiled Geophysical Prospecting Surveys in Groundwater Research in the Metropolitan District of Quito in Northern Ecuador" [7] compiles and examines 23 geophysical prospecting surveys of interest in groundwater research in the Metropolitan District of Quito, including 7 ERT, 8 VES, 4 REMI and 1 FDEM surveys for shallow Holocene and late Pleistocene formations, and 3 L-MTS surveys for Holocene to late Pliocene formations. No surveys exploring the complete saturated thickness of the Pliocene aquifers could be compiled. This gap is impeding the assessment of the groundwater fraction of these regional aquifers that can be exploited sustainably.

The paper "Geophysical Characterization in the Shallow Water Estuarine Lakes of the Southern Everglades, Florida" [8] uses FER and TDEM techniques to understand the spatiotemporal variations of surface water and shallow groundwater salinity in the coastal lakes of the Everglades National Park (ENP) in south Florida in southeast USA. Anthropogenic activities have altered freshwater flows through ENP, such that saltwater has intruded inland from the coastline, causing coastal lakes and their ecosystems to be exposed to higher salinity conditions. Geophysical surveys assessed the spatiotemporal distribution of salinity needed to evaluate restoration efforts.

3. Conclusions

The Guest Editors envision that the published papers in this Special Issue would be of interest to researchers and practitioners, and help identify further research initiatives. We also hope that the readers can find the material of this Special Issue both interesting and inspiring when exploring geophysical methods for shallow aquifers characterization and modeling. The findings and techniques presented in this collection of papers contribute to the increasing interest in groundwater research.

Funding: This research received no external funding.

Acknowledgments: The authors of this paper, who served as the Guest Editors of this Special Issue, wish to thank the journal editors, all authors submitting papers, and the referees who contributed to revising and improving the eight published papers.

Conflicts of Interest: The authors declare no conflict of interest.

References

1. Paz, M.C.; Alcalá, F.J.; Medeiros, A.; Martínez-Pagán, P.; Pérez-Cuevas, J.; Ribeiro, L. Integrated MASW and ERT imaging for geological definition of an unconfined alluvial aquifer sustaining a coastal groundwater-dependent ecosystem in southwest Portugal. *Appl. Sci.* **2020**, *10*, 5905. [CrossRef]
2. Rey, J.; Martínez, J.; Mendoza, R.; Sandoval, S.; Tarasov, V.; Kaminsky, A.; Hidalgo, M.C.; Morales, K. Geophysical Characterization of Aquifers in Southeast Spain Using ERT, TDEM, and Vertical Seismic Reflection. *Appl. Sci.* **2020**, *10*, 7365. [CrossRef]
3. Sendrós, A.; Urruela, A.; Himi, M.; Alonso, C.; Lovera, R.; Tapias, J.C.; Rivero, L.; Garcia-Artigas, R.; Casas, A. Characterization of a Shallow Coastal Aquifer in the Framework of a Subsurface Storage and Soil Aquifer Treatment Project Using Electrical Resistivity Tomography (Port de la Selva, Spain). *Appl. Sci.* **2021**, *11*, 2448. [CrossRef]
4. Martínez-Segura, M.A.; Conesa-García, C.; Pérez-Cutillas, P.; Martínez-Pagán, P.; Vásconez-Maza, M.D. Identifying Changes in Sediment Texture along an Ephemeral Gravel-Bed Stream Using Electrical Resistivity Tomography 2D and 3D. *Appl. Sci.* **2021**, *11*, 3030. [CrossRef]

5. Alcalá, F.J.; Martínez-Pagán, P.; Paz, M.C.; Navarro, M.; Pérez-Cuevas, J.; Domingo, F. Combining of MASW and GPR Imaging and Hydrogeological Surveys for the Groundwater Resource Evaluation in a Coastal Urban Area in Southern Spain. *Appl. Sci.* **2021**, *11*, 3154. [CrossRef]
6. Marques, T.; Matias, M.S.; Ferreira da Silva, E.; Durães, N.; Patinha, C. Temporal and Spatial Groundwater Contamination Assessment Using Geophysical and Hydrochemical Methods: The Industrial Chemical Complex of Estarreja (Portugal) Case Study. *Appl. Sci.* **2021**, *11*, 6732. [CrossRef]
7. Peñafiel, L.; Alcalá, F.J.; Senent-Aparicio, J. Usefulness of Compiled Geophysical Prospecting Surveys in Groundwater Research in the Metropolitan District of Quito in Northern Ecuador. *Appl. Sci.* **2021**, *11*, 11144. [CrossRef]
8. Kiflai, M.E.; Whitman, D.; Price, R.M.; Frankovich, T.A.; Madden, C.J. Geophysical Characterization in the Shallow Water Estuarine Lakes of the Southern Everglades, Florida. *Appl. Sci.* **2022**, *12*, 1154. [CrossRef]

applied sciences

MDPI

Article

Integrated MASW and ERT Imaging for Geological Definition of an Unconfined Alluvial Aquifer Sustaining a Coastal Groundwater-Dependent Ecosystem in Southwest Portugal

Maria Catarina Paz [1,2,*], Francisco Javier Alcalá [3,4], Ana Medeiros [5], Pedro Martínez-Pagán [6], Jaruselsky Pérez-Cuevas [7] and Luís Ribeiro [5,†]

1 CIQuiBio, Barreiro School of Technology, Polytechnic Institute of Setúbal, 2839-001 Lavradio, Portugal
2 Instituto Dom Luiz, Faculdade de Ciências, Universidade de Lisboa, 1749-016 Lisbon, Portugal
3 Instituto Geológico y Minero de España (IGME), 28003 Madrid, Spain; fj.alcala@igme.es
4 Instituto de Ciencias Químicas Aplicadas, Facultad de Ingeniería, Universidad Autónoma de Chile, Santiago 7500138, Chile
5 Civil Engineering Research and Innovation for Sustainability (CERIS), Instituto Superior Técnico, Universidade de Lisboa, 1049-001 Lisbon, Portugal; anamedeiros@tecnico.ulisboa.pt (A.M.)
6 Department of Mining and Civil Engineering, Universidad Politécnica de Cartagena, 30203 Cartagena, Spain; p.martinez@upct.es
7 Engineering Sciences, Pontificia Universidad Católica Madre y Maestra, Santo Domingo 2748, Dominican Republic; jaruselskyperez@pucmm.edu.do
* Correspondence: catarina.paz@estbarreiro.ips.pt
† Deceased.

Received: 6 July 2020; Accepted: 22 August 2020; Published: 26 August 2020

Abstract: This paper integrates multichannel analysis of surface waves (MASW) and time-lapse electrical resistivity tomography (ERT) to define aquifer geometry and identify transient groundwater features of the Cascalheira Stream Basin Holocene alluvial aquifer (aquifer H), which contributes to the Santo André Lagoon, part of a coastal groundwater-dependent ecosystem (GDE), located in southwest Portugal. MASW measures shear-wave velocity (VS), allowing one to obtain steady geological models of the subsurface, and ERT measures subsurface electrical resistivity (ER), being subjected to ambient changes. MASW enables disambiguation of geological structures in low ER environments, such as coastal areas. This research covered one natural year and involved one MASW campaign, four ERT campaigns, and additional geological field surveys and groundwater monitoring to assist interpretation of results. In the area, the conjugate NW–SE and NE–SW strike-slip fault systems determine compartmentalization of geological structures and subsequent accommodation space for Holocene sedimentation. MASW and ERT surveys show how the NW–SE system deepens these structures toward the coast, whereas the NE–SW system generates small horsts and grabens, being one of these occupied by aquifer H. From upstream to downstream, aquifer H thickness and width increase from 10 m to 12 m and from 140 m to 240 m, respectively. Performance of VS and ER models was satisfactory, with a normalized error of the VR and ER models in the 0.01–0.09 range, meaning that a quantitative quota of uncertainty can be segregated from the overall uncertainty of groundwater models without substantially affecting its simulations accuracy. This methodology seeks to improve the design of shallow groundwater research in GDE preservation policies.

Keywords: multichannel analysis of surface waves; electrical resistivity tomography; time-lapse inversion; aquifer geometry; groundwater-dependent ecosystem; Santo André Lagoon; Portugal

1. Introduction

Aquifers play a critical role in sustaining the economy and the environment of coastal areas [1]. These areas are often subjected to high rates of groundwater drawing to meet the increasing urban, tourism, industrial, and agricultural demands, adding stress to groundwater bodies and dependent ecosystems [2,3]. The combination of global climate forces underlying human pressures threatens the fragile balance between freshwater and saltwater and therefore the quantity and quality levels required for a good functioning of groundwater-dependent ecosystems (GDE) [4]. These effects are especially visible in small unconfined aquifers because their greater exposure to human impacts and smaller storage capacity may limit a regular water provision to ecosystems during drier periods [5,6].

This is the case of the Santo André Lagoon (SAL), which, together with the Sancha Lagoon, form a coastal GDE space in southwest Portugal whose hydrological functioning depends on regular inputs of fresh surface water and groundwater [7,8]. This GDE space was catalogued in 1993 as a Special Protection RAMSAR Area due to its ecological value for wildlife preservation [RAMSAR website: https://www.ramsar.org/es/humedal/portugal]. This type of ecosystem is environmentally protected by the European Water Framework Directive [9], which also establishes the obligation to permanently characterize its hydrological functioning.

Preliminary findings from the Portuguese R&D Groundscene Project [10,11] showed that freshwater contribution from tributary streams, which in turn are groundwater-dependent, regulates the water salinity balance of the SAL ecosystem. The Groundscene Project left a blank in what concerns the detailed hydrological functioning needed for predictive modeling under scenarios of climate change and subsequent new land and water uses for human adaptation. This is a large task in progress that is being boarded through implementing and coupling well suited rainfall-runoff and groundwater modeling tools [12–14].

Over the base of detailed and existing or compiled weather, land use, and water fluxes datasets, implementation of groundwater modeling tools involves three general stages for aquifer conceptualization [15–17]: (1) geometry definition, (2) acquisition of hydraulics data, and (3) evaluation of water balance components. This paper is aimed at advancing the first stage, i.e., aquifer geometry definition, and at qualifying some transient groundwater features subordinately. In flat coastal areas, near-surface geophysical techniques have been widely used in groundwater research to acquire this basic information [8,18–24]. These techniques are non-invasive, usually cheap to apply, and useful when geotechnical sounding data are sparse or not able to provide detailed subsurface information required for groundwater modeling, such as aquifer geometry and some transient groundwater features.

This paper integrates multichannel analysis of surface waves (MASW) and time-lapse electrical resistivity tomography (ERT) to define aquifer geometry and qualitatively identify some transient groundwater features of the alluvial aquifer contributing to the SAL. Integrated MASW and ERT results were intentional for this purpose because the MASW technique responds to the steady shear modulus of subsurface materials, expressing seismic shear-wave velocity (VS) [25–28], whereas ERT responds to the electrical conductivity (EC) of subsurface media, which is subjected to transient ambient changes [21,29–31].

The use of MASW in groundwater research is incipient [32–34] and has mostly addressed disambiguation of the subsurface geological structures deduced from ERT in high EC environments, such as coastal areas [33,35–37]. MASW and ERT show advantages and limitations in what concerns the reached exploration depth and the resolution of the subsurface geological structures. MASW provides higher exploration depths than ERT but less detailed resolution, whereas ERT is highly responsive to detailed subsurface EC changes related to geological and hydrological heterogeneities [23,29,31,38]. The limited capability of ERT for geological definition contrasts with its ability to identify temporal groundwater changes. This way, time-lapse ERT can be used to disambiguate steady features of geological media (e.g., bulk density, clay content, and organic matter content) from transient features of saturated media (e.g., groundwater storage, dissolved ions). These transient features result from the variable combination of the non-evaporative fractions of precipitation [6] and atmospheric salinity rates [39] reaching the water table over time.

This research covered the natural year 2014 and involved a geophysical survey characterized by one MASW campaign and four (quarterly) ERT campaigns. Besides this, two other complementary tasks were: a geological survey to define the local faulting and its correspondence to the regional structural setting; and aquifer monitoring, with quarterly piezometry measurements to draw water table fluctuation along the preferential groundwater flow path, and some EC measurements to support the hydrogeological interpretation of ERT surveys. Since the contributing area to the SAL is too large for making up these tasks in detail, only the Cascalheira Stream Basin (CSB), the largest and most representative tributary, was selected to define the geological structure and the preliminary hydrological functioning of the Upper Miocene—Quaternary sedimentary body hydraulically connected to the SAL. For a fluent reading, a description of acronyms used is in Table 1.

Table 1. Description of acronyms used.

Acronym	Definition
Aquifer H	Holocene alluvial aquifer
CSB	Cascalheira Stream Basin
CVMAE	Normalized MAE
CVRMSE	Normalized RMSE
CVSTD	Normalized STD
EC	Electrical conductivity
ER	Electrical resistivity
ERT	Electrical resistivity tomography
GDE	Groundwater-dependent ecosystem
GEC	Groundwater electrical conductivity
GER	Groundwater electrical resistivity
lnNSE	Logarithmic form of NSE
MAE	Mean error
MASW	Multichannel analysis of surface waves
MRE	Mean relative error
NAO	North Atlantic Oscillation
NSE	Nash–Sutcliffe efficiency coefficient
PBIAS	Percent bias
R^2	Coefficient of determination
RD	Relative difference
RMSE	Root-mean-square error
RSR	RMSE relative to STD
SAL	Santo André Lagoon
STD	Standard deviation of the measured data
SWQM	SAL water quality monitoring
VS	Shear-wave velocity
WFD	European Water Framework Directive

2. Study Area

2.1. Location and Climate

The study area is located at the outlet of the CSB (08°38′–08°47′ W, 38°05′–38°07′ N) in the southwest coast of Portugal (Figure 1a). The CSB covers an area of 31.5 km^2, has a mean elevation of 230 m a.s.l. (outlet is 2 m a.s.l. on the West and peak elevation is 290 m a.s.l. on the East), and its surface water and groundwater components flow westwards to the SAL [40]. The CSB is the main tributary to the SAL.

Climate is warm-summer Mediterranean according to the Köppen classification, which means temperate dry summers and rainy winters [11,41]. Weather data compiled from Sines and Monte Velho meteorological stations (Figure 1a) show that precipitation (P) occurs in three distinctive phases: (1) a predominant rainy phase from October to January with average monthly P from 70 to 90 mm, which represents around 60% of annual P; (2) a moderately rainy phase from February to May with average monthly P from 30 to 50 mm, which means around 30% of annual P; and (3) a dry phase from

June to September with average monthly P below 10 mm and occasional rainfall events exceeding 30 mm [40]. Average annual P is around 505 mm with a coefficient of variation of 0.35 over the period 1970–2016. Average monthly temperature (T) varies from 10.6 °C in January to 19.5 °C in July. Average minimum and maximum monthly T vary in the 5.8–15.4 °C and 14.1–24.9 °C ranges in January and July, respectively [42]. Average annual T is around 15.1 °C with a coefficient of variation of 0.05 over the period 1973–2007 [40]. Average annual actual evapotranspiration (E) is around 410 mm, thus the E-to-P ratio is about 0.2 [42].

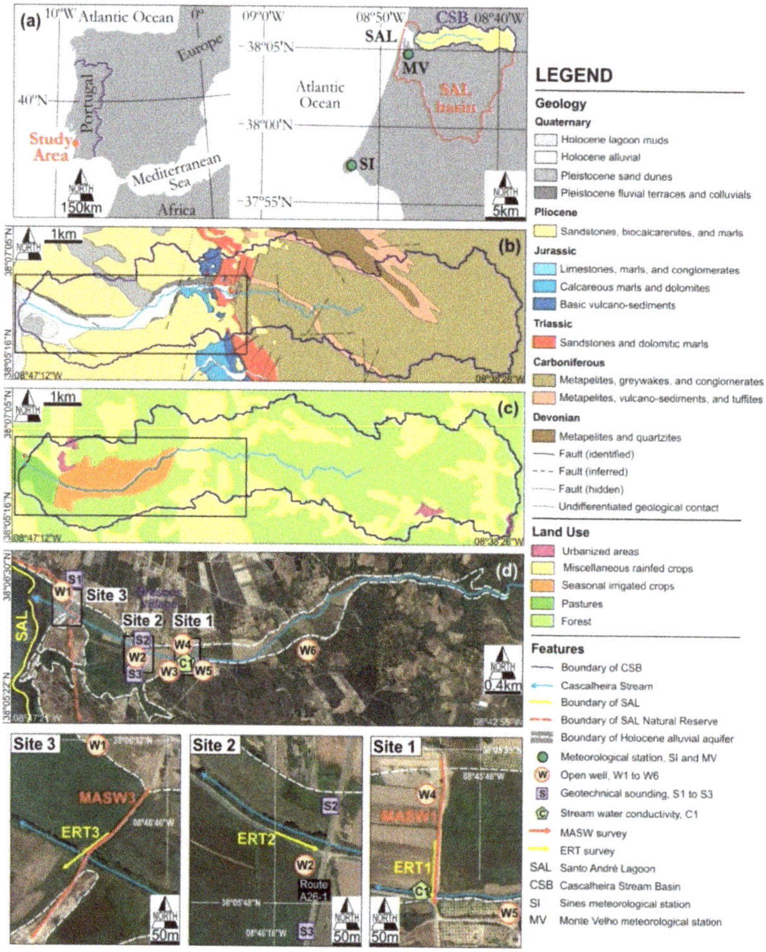

Figure 1. (**a**) Location of the study area in southwest Portugal, showing the Santo André Lagoon (SAL) and its contributing basin, the Cascalheira Stream Basin (CSB), and the Sines (SI) and Monte Velho (MV) meteorological stations. (**b**) Geological map (scale 1:50,000) of the CSB according to [43] and direct field observations. (**c**) Land-use units of the CSB according to [42], aerial photographs, and direct field observations. (**d**) Holocene alluvial aquifer (aquifer H) contributing to the SAL, showing geophysical surveys performed along the main groundwater flow path at sites 1 (MASW1, ERT1), 2 (ERT2), and 3 (MASW3, ERT3), aquifer monitoring points (handmade open wells W1 to W6), geotechnical soundings S1 to S3 after [40], stream water EC measurements within the SAL Water Quality Monitoring (SWQM) Programme, and geographical features and sites cited in the text.

2.2. Geological and Hydrogeological Setting

The study area (Figure 1b) belongs to the western Mesozoic extensional margin of the Iberian Plate, which has had a long distensive tectonic and magmatic history from the Late Triassic up to the present [44,45]. Locally, the CSB includes the following synthetic succession from bottom to top [40,46–49]: (1) Variscan crystalline bedrock includes (i) Devonian low-grade metapelites and quartzites, and (ii) Carboniferous low-grade metapelites, vulcano-sediments, and tuffites underlying to metapelites, greywakes, and conglomerates; (2) Mesozoic sedimentary cover includes (i) Triassic continental sandstones intruded by basaltic and doleritic dikes, and (ii) Lower Jurassic platform dolomites, carbonates, and marls; and (3) Upper Miocene to present sedimentary filling includes (i) Pliocene marine sandstones, biocalcarenites, and marls, (ii) Pleistocene continental fluvial terraces and clay-rich colluvials underlying to coastal sand dunes, and (iii) Holocene alluvial, lagoon, and beach sands sediments.

In this extensional geological domain, the conjugate NE–SW and NW–SE strike-slip fault systems affecting underlying formations since the Neogene determine the accommodation space for Holocene sedimentation [43,47], the subject of this paper. As a result, the Holocene sedimentary record may reach 20 m in thickness in the CSB–SAL boundary (Figure 1b) with the following detailed succession from bottom to top (1) alternating marine muds and tidal sands below the SAL, (2) organic matter-rich lagoon muds with more than 20% of bioclasts below the SAL, (3) fluvial coarse sands and sandy loam with some bioclasts in the old (currently inland) lagoon space, and (4) coarse-grained deposits occupying the current flood plain of the Cascalheira stream valley [40,46–48].

In hydrogeological terms, geological formations can be classified into four groups according to the permeability type and the storage capacity reported by [46,50]: (1) Devonian and Carboniferous metapelitics are low-permeability formations forming the impervious basement of local aquifers; (2) Triassic sandstones and Lower Jurassic carbonates form moderately to highly permeable aquifers with karst features and thickness up to 200 m; (3) Pliocene and Pleistocene sediments are low-to-moderate permeable formations forming an unconfined multilayer aquifer of about 30 m in thickness hydraulically connected to the uppermost Holocene formations; and (4) Holocene alluvial deposits comprise sand, gravels, and silt of 10 m to 20 m in thickness filling the Cascalheira Stream valley, an unconfined aquifer (hereafter aquifer H) of 1.46 km^2 at the CSB outlet that contributes to the SAL (Figure 1b).

The hydrogeological functioning of the CSB depends on water exchange between aquifer H and underlying Pliocene and Pleistocene formations, the extent of weathered metapelitics slopes upstream, and thickness and extent of aquifer H downstream (Figure 2). Recharge to aquifer H comes from direct rainfall and runoff infiltration, interflow runoff from fissured and weathered metapelitics upstream, transferences from underlying formations, and some irrigation return from agriculture. Discharge from aquifer H depends on actual evapotranspiration and the scarce groundwater pumping in the area. The resulting positive net groundwater balance from aquifer H sustains the freshwater-saltwater balance of the SAL. However, groundwater contribution to the SAL is temporally retained at the aquifer H outlet because the existence of low-permeability (clay-rich) interbedded lagoon sediments diminishes the hydraulic connectivity between aquifer H and the SAL, and saltwater-freshwater interface within the sediments underlying the SAL forces fresh groundwater to be discharged above it [7,51]. The result is that the piezometric level of aquifer H rises above topography after rainfall-recharge events and a significant fraction of groundwater flooding the CSB outlet is evacuated to the SAL as surface runoff [7,8]. Additionally to this natural functioning, groundwater renovation depends on the annual opening of the coastal sand bar that separates the SAL from the Atlantic Ocean to favor discharge of temporary surface water retained at the aquifer H outlet and the subsequent groundwater level depletion required for agricultural and livestock practices inland [11]. In 2014, the coastal sandbar was opened on 28 February.

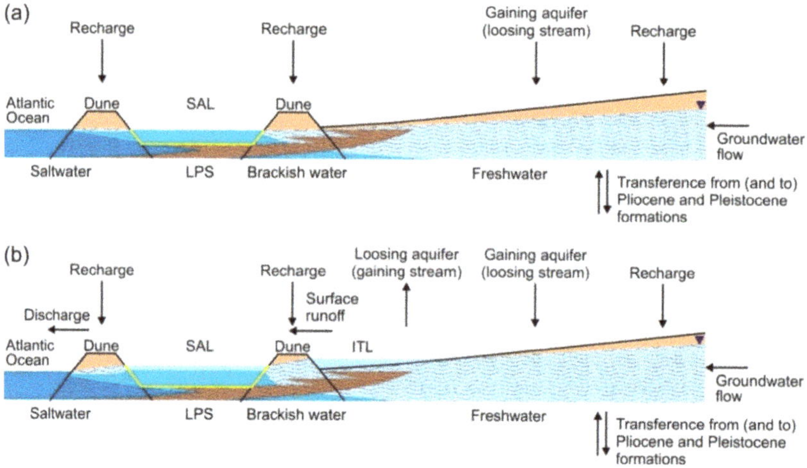

Figure 2. Conceptualization of the hydrogeological functioning of aquifer H during (**a**) the dry phase in summer and (**b**) predominant rainy phase in winter. Scale-out scheme after [7,46,51], and direct field observations. SAL, Santo André lagoon; LPS, low-permeability (clay-rich) lagoon sediments; ITL, inland temporary lagoon.

2.3. Land and Groundwater Use

In the CSB, land use is mostly devoted to different rainfed crops and forest, 7% is irrigated crops, 5% is pasture, and less than 1% is urbanized areas (Figure 1c). Irrigated crops (mainly corn) are practically restricted to the aquifer H floodplain during spring season, after the coastal sandbar is opened. In this irrigated land, the average soil texture is 53% sand, 31% silt, 14% clay, and 4% organic matter, with maximum soil depth reaching up to 1.3 m. Water allocation for irrigation is around 6600 m^3 per hectare and year [42]. Irrigation water comes from upstream derivations, so the existing hand-made open wells are scarcely used. Irrigation is done through drip systems, which reduce evaporation and infiltration losses and increases salinity of irrigation return. At the west end of the study area, the SAL and some neighboring spaces (Figure 1d) were catalogued in 1993 as a Special Protection RAMSAR Area due to its ecological value for wildlife preservation [RAMSAR website: https://www.ramsar.org/wetland/portugal].

3. Methods

3.1. Overall Framework for Data Collection

Frequency for geophysical surveying and subsequent groundwater monitoring was adapted to the particular groundwater quantity (flow) and quality (EC) temporal dynamics of aquifer H. In small, high-yielding alluvial aquifers having almost null groundwater exploitation such as aquifer H, the average groundwater turnover time identifies well the infilling–emptying cycles produced from rainy and dry phases, and consequently the minimum monitoring frequency required to characterize groundwater flow and EC responses. The average groundwater turnover time was deduced from preliminary aquifer geometry and hydraulics data by assuming a predominant piston (plug) flow condition [5,6,52] as:

$$G = b \cdot A \cdot S / T, \tag{1}$$

where G is average groundwater turnover time [T], b is a dimensionless flow-path form parameter determined by the aquifer geometry, A is aquifer surface [L^2], S is the dimensionless aquifer storage coefficient or drainable porosity for unconfined aquifers, and T is aquifer transmissibility [L^2 T^{-1}], which is the product of aquifer hydraulic conductivity [L T^{-1}] and saturated thickness [L].

Aimed at interpreting the ERT surveys properly [23], aquifer monitoring included piezometry and EC measurements in selected handmade open wells (Figure 1d). Piezometry was measured using a level probe from Seba Hydrometrie, with a precision of 0.005 m. EC was measured each 1-m from 1 m below water table to 8 m depth using a multi parameter probe from Hanna Instruments, with a precision of 0.01 mS cm^{-1}; the average value of these measures was calculated for each well.

3.2. MASW Surveys

MASW is a seismic geophysical technique in which the Rayleigh wave fundamental mode dispersion curve and higher modes (if present) are extracted from a shot record and then inverted to generate a 1D VS [L T^{-1}] model [27,28]. This technique allows for analyzing the fundamental and higher modes simultaneously, thus permitting to obtain more accurate VS models [25,26]. A roll-along setup with a land-streamer acquisition system was used to obtain a continuous 2D VS model. This procedure enables us to acquire data rapidly because it is not necessary to plant the geophones each time a measurement is made.

MASW data were acquired using a 24-channel SUMMIT II Compact Seismograph by DMT, Germany, with the following configuration: recording array of 24 vertical component geophones, 2-m geophone spacing, 4-m separation between the source impact point and first geophone to minimize near-source effects, two stacks, 10-m displacement between readings, and a sampling rate of 0.25 m s. A 5 kg sledge hammer was used to generate the Rayleigh waves.

Data analysis was carried out with SurfSeis3 software® by the Kansas Geological Survey, The University of Kansas, USA. Data processing consisted of geometry edition, data filtering, muting (when needed), generation of overtones (frequency–time energy diagrams), and fundamental and higher modes (if present) identification. Finally, dispersion curves were determined and then subjected to a mathematical inversion process to obtain continuous 2D VS models. These were plotted using the triangulation with linear interpolation method, which gives good results for evenly distributed data over the mapping area.

3.3. ERT Surveys

ERT is an electrical geophysical technique that uses measurements of voltage between two reading electrodes installed in land surface, once direct current in two other electrodes is injected. This technique allows for calculating the subsurface electrical resistivity (ER) [Ω m], reciprocal of EC [S m^{-1}] [29,31,38], as:

$$ER = 1/EC, \quad (2)$$

Disposition of the electrodes changes depending on the array used, so a grid of subsurface apparent resistivity values is obtained. These values are then mathematically inverted to obtain subsurface ER models. Penetration depth depends on subsurface EC, which is a function of pore-water EC and ground EC, the input voltage used, and the electrode spacing adopted [38].

ERT data were acquired using a GL-16 resistivity meter with a P-100-2 accumulator ® by PASI Instruments. A dipole–dipole array was the electrode disposition used since it provides good resolution both on vertical and horizontal directions. Configuration was: 6-m electrode spacing, 36-m maximum separation between dipoles, and 200 V as the input voltage applied.

Since the resistivity meter used is not automatic, it was possible to repeat each measurement in the field, whenever the data represented an outlier to the dataset, guaranteeing the acquisition of good quality data. Data were preprocessed in order to calculate the apparent resistivity measured in each node of the subsurface grid. Simultaneous space–time inversion of time-lapse ERT data was carried out using the RES2DINV software® by Geotomo Software, with a blocky constraint to minimize exaggeration of smooth model changes when subsurface changes are locally limited [53] and a severe reduction of side blocks effects to minimize exaggeration from robust inversion [54]. Iterations were limited to three so as not to create artifacts, since the inversion error had low reduction in third iteration comparatively to former iterations. In profiles ERT1 and ERT3, the electrode spacing unit was reduced by half (3 m) to minimize the effect of large near-surface resistivity variations, as proposed by [54]. Similarly to VS models, ER models were also plotted using the triangulation with linear interpolation method.

3.4. Topographic Correction of 2D VS and ER Models

Topographic correction of 2D VS and ER models followed two steps. First, ground relief was measured in the field using a leveled civil-work laser and a vertical leveled ruler. Height regarding the ground was measured in equally spaced points along profiles. The first profile point was considered the relative vertical zero from which the other measured heights were summed or subtracted to create relative relief profiles. Later, relative relief profiles were converted into georeferenced elevation profiles using the Earth Digital Model from the ArcGIS software® by ESRI.

4. Results

4.1. Frequency for Geophysical Surveying

Frequency for geophysical surveying and subsequent groundwater monitoring was calculated through Equation (1). Geological information deduced from geotechnical soundings S1 to S3 (Figure 1d) after [40] and hydrogeological data compiled from [7,10] were used for this attempt. Taking average $b = 1.1$ for predominant longitudinal groundwater flow paths [55] such as in aquifer H, $A = 1.46$ km^2 (Figure 1b), $S = 0.05$ [40], and $T = 300$ m^2 day^{-1} and 800 m^2 day^{-1}, measured respectively in dry and rainy phases by [10], and G varied in the 3.2–8.9-month range.

All geophysical surveys were performed during 2014 (Figure 3). Long-term significance of the selected year 2014 was evaluated from the analysis of the global North Atlantic Oscillation index [NAO website: http://www.cpc.ncep.noaa.gov/] (Figure 3a). The NAO index controls long-term precipitation and temperature regimes in southern Portugal [56], and therefore actual evapotranspiration (E) and stream flow rates [41]. The selected year 2014 fits well to the average hydrological condition deduced from the long-term P (Figure 3b) and E (Figure 3c) time series from Sines meteorological station (Figure 1a). Thus, geophysical and hydrogeological interpretations should be framed into the context of an average hydrological condition in the area.

In this context, three months was the optimal frequency adopted for ERT surveying (Table 2) and subsequent groundwater monitoring (Table 3) at sites 1, 2, and 3 (Figure 1d). This frequency covers adequately the about four-month long predominately rainy, moderately rainy, and dry phases taking place in the area, as shown graphically in Figure 4. Only one MASW survey was performed at sites 1 and 3 (Table 2).

Table 2. Description of MASW and ERT surveys.

Site	Profile ID [1]	Length, m	Prospecting Depth, m	Date
1	MASW1 ERT1	230 90	30 13	23 June 2014 13 March 2014 4 June 2014 12 September 2014 10 December 2014
2	ERT2	78	15	13 March 2014 4 June 2014 12 September 2014 10 December 2014
3	MASW3 ERT3	310 108	27 13	23 June 2014 12 March 2014 3 June 2014 10 September 2014

[1] ID and location as in Figure 1d.

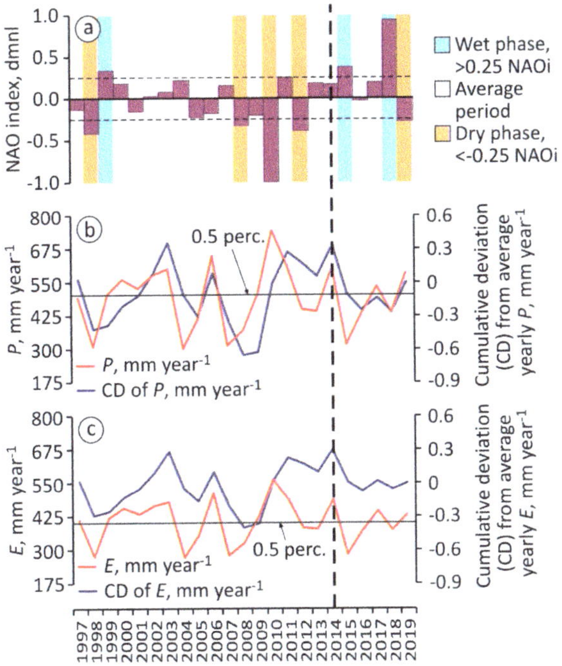

Figure 3. For natural years 1997–2019 in the SAL area, (**a**) normalized North Atlantic Oscillation (NAO) index (purple bars) [NAO website: http://www.cpc.ncep.noaa.gov/]; (**b**) annual precipitation (P) time series from Sines meteorological station and cumulative deviation (CD) from average annual P, mm year^{-1}; and (c) annual actual evapotranspiration (E) time series from Sines meteorological station and CD from average annual E, mm year^{-1}. For (**b**,**c**), the average (0.5 percentile) of yearly data series is indicated. Vertical dotted line indicates the study year 2014.

Table 3. Description of monitored variables in open wells.

Site	ID [1]	Elevation, m a.s.l.	Aquifer and Flow Zone	Variable [2]	GEC [3]	GER [4]
Upstream	W6	35.18	Pliocene, recharge	PL, GEC	200	50
1	W5	14.06	Pleistocene, transit	PL, GEC	500	20
	W3	9.00	Pleistocene, discharge	PL		
	W4	10.07	Holocene, recharge	PL		
2	W2	8.82	Holocene, transit	PL		
3	W1	4.57	Pliocene, discharge	PL, GEC	393	25

[1] ID and location as in Figure 1d. [2] PL is piezometric level as in Figure 4. [3] GEC is groundwater electrical conductivity in µS cm^{-1} measured on 20 August 2014 when the multi-parameter probe was available; wells W3 and W4 were not accessible. [4] GER is groundwater electrical resistivity in Ω m after GEC reversion using Equation (2).

4.2. Hydrogeophysical Basis for VS and ER Models Interpretation

When MASW and ERT are aimed at defining geological structures in porous saturated media, a preliminary conceptualization of properties governing the magnitude of variables VS (MASW) and ER (ERT) is needed. Some basic interpretative criteria are described below.

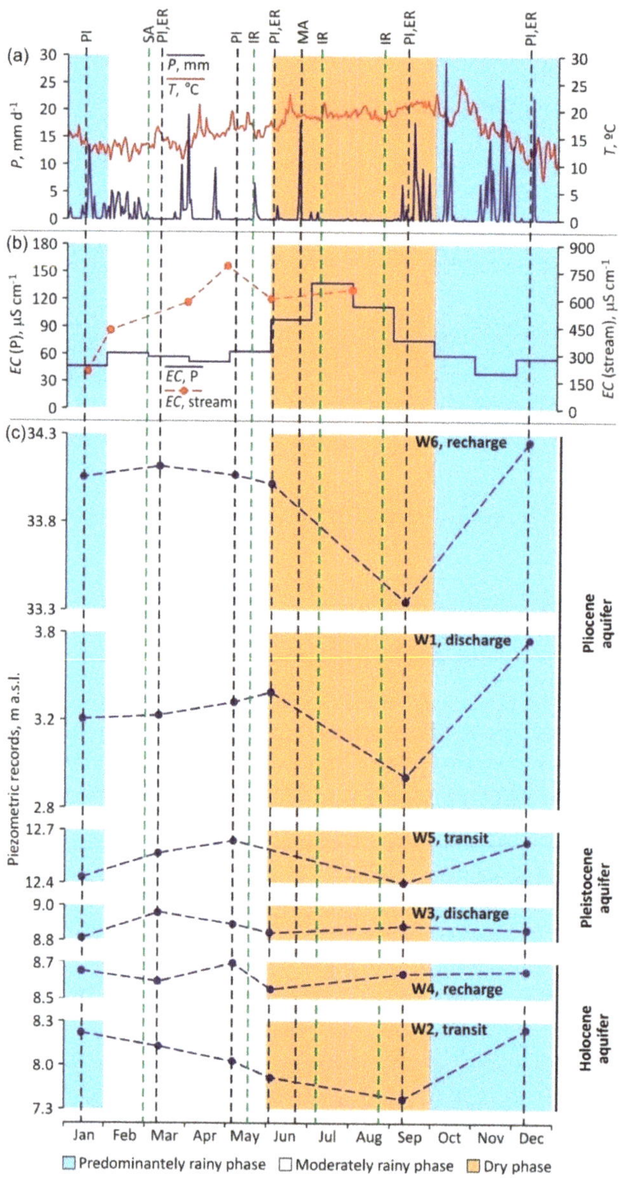

Figure 4. For year 2014 in the SAL area, (**a**) 24-h P and T distribution, after P (mm) and T (°C) records from Sines meteorological station (Figure 1a); (**b**) monthly distribution of average precipitation-weighed EC after EC (µS cm^{-1}) and P (mm) records from Monte Velho meteorological station (Figure 1a), and some Cascalheira stream water EC (µS cm^{-1}) measurements at site 2 after the SWQM Programme (Figure 1d); and (**c**) piezometry (m a.s.l.) in handmade open wells W1 to W6 (Table 3) ordered according to the monitored geological formation (Pliocene, Pleistocene, Holocene) and groundwater flow zone (recharge, transit, discharge). Vertical black dotted lines indicate geophysical surveying (Table 2) and groundwater monitoring (Table 3) as ERT (ER) and MASW (MA) surveys, and piezometry (PI). Vertical green dotted lines indicate human actions modifying water dynamics as SAL aperture (SA) and groundwater pumping for irrigation (IR).

In soft coarse-grained sediments, VS propagation is considered a site-specific steady property determined by effective compaction, and as such, it is dependent on the age and depth of each geological material piled on vertical [57,58]. Table 4 summarizes some reference VS values for soft and stiff geological materials equivalent to those described in the SAL area [34,59]. Different relationships between VS and age and depth for other similar lithologies can be found in the scientific literature [60–64], which can be used to enlarge the ranges reported in Table 4. For the sedimentary formations in the study area, VS values in the 50–900 m s^{-1} range are expected.

Table 4. Summary of some reference VS ranges compiled from the scientific literature for different sediments and rocks, and its equivalence to the geological materials described in the SAL area.

Geomaterial	VS, m s^{-1}	Reference	Equivalence [1]
Soft clay	80–200	[34]	Holocene clay
Loose sand	80–250	[34]	Holocene sand
Loose sand and gravel	100–200	[59]	Holocene sand and gravel
Anthropogenic filling	50–100	[59]	Holocene floodplain
Cropland and organic soil	50–150	[59]	Holocene floodplain
Stiff clay	200–600	[34]	Pleistocene clay
Dense sand	150–500	[34]	Pleistocene sand dunes
Soft-stiff sand	300–500	[59]	Pleistocene sand
Stiff gravel	300–600	[34]	Pleistocene conglomerate
Cemented clay	600–1000	[59]	Pliocene marl
Cemented sand	500–900	[59]	Pliocene calcarenite
Cemented gravel	500–900	[34]	Pliocene conglomerate
Weathered carbonate bedrock	600–1000	[34]	Jurassic marls
Weathered crystalline bedrock	800–1200	[59]	Variscan weathered metapelites
Hard carbonate bedrock	1200–2500	[34]	Jurassic carbonates
Hard crystalline bedrock	1500–2500	[59]	Variscan metapelites

[1] Geological description and location in Figure 1c.

In porous saturated media, ER corresponds to EC inferred from certain intrinsic geological features and groundwater salinity; the former remains unvaried, whereas the latter varies over time. In pristine coastal areas, groundwater electrical resistivity (GRE) determines ER when salinity contribution from subsurface geology is negligible. In this case, in time-lapse ER models, transient GER changes ultimately depend on the variable combination of the non-evaporative fractions of precipitation and atmospheric bulk deposition salinity reaching the water table over time, which are predictable, measurable variables. For instance, atmospheric bulk deposition EC decreases inland, so inland GEC (recharge zone) is typically less than coastal GEC [5,39]. How the GER changes extend over space and time is crucial to define the hydraulic boundaries of aquifer H.

In the SAL coastal fringe, atmospheric bulk deposition EC was deduced from the long-term Monte Velho meteorological station (Figure 1a), which belongs to the *Co-operative Programme for Monitoring and Evaluation of the Long-Range Transmission of Air Pollutants in Europe* [EMEP network: http://www.emep.int/] aimed at providing chemical analysis of precipitation [5,39]. In year 2014, monthly precipitation-weighed EC varied in the 39–138 µS cm^{-1} range with typical values of 40–80 µS cm^{-1} in the rainy phase and higher than 100 µS cm^{-1} in the dry phase (Figure 4b). The values in the rainy phase determine the expected GEC in aquifer H since aquifer recharge from P in the dry phase is negligible (Figure 4a). Applying the monthly E-to-P ratios deduced from daily P and T time series (Figure 3c) to the atmospheric bulk deposition EC in the rainy phases, monthly GEC results in the 200–400 µS cm^{-1} range; GER being in 25–50 Ω m range using Equation (2). In addition to this theoretical appraisal, GEC was measured in pristine (without apparent human influence) Pliocene (W1 and W6) and Pleistocene (W5) wells to corroborate this regional GEC baseline (Table 3). GEC varied in the 200–500 µS cm^{-1} range, which is similar to the theoretical range described above, thus GER being in 20–50 Ω m range. Unfortunately, there are no inland atmospheric bulk deposition EC data, although values 0.2-fold of the magnitude reported in the coastal fringe can be assumed after the decreasing inland gradient documented by [39]. Applying this decreasing gradient to the theoretical coastal GEC and GER, inland GEC and GER can be tentatively proposed in the 40–100 μS cm^{-1} and 100–250 Ω m ranges, respectively. This theoretical GER baseline is a reference for interpreting experimental ER models in the study area.

Piezometry in wells W1 to W6 enables us to deduce how rainy phases and human actions determine GER changes in each geological formation (Figure 4; Table 3). This hydrodynamics information is crucial to qualify transient GER changes in sequential time-lapse ERT. Piezometry in Pliocene wells W1 and W6 follows the seasonal weather phases (Figure 4c). In the recharge zone (W6), delayed responses of about two months and three months during the rainy and dry phases, respectively, are observed. In the discharge zone (W1), piezometry lightly depletes after the SAL opened on 28 February. Piezometry in Pleistocene wells W3 and W5 coarsely follows the weather phases with a delayed response of about three months during both rainy and dry phases in the transit zone (W5); there is no data for the recharge zone (Figure 4c). In the discharge zone (W3), the lightly influence of groundwater pumping after the rainy phase and of irrigation return after the dry phase are observed; note that W3 is hydraulically connected to aquifer H. Piezometry in Holocene wells W2 and W4 is especially influenced by human actions (Figure 4c). In the recharge zone (W4), the effects of SAL opening (with depletion in March), groundwater pumping for irrigation (with depletion in June), and return of a fraction of the imported irrigation water (with rising in September) are observed. In the transit zone (W2), piezometry progressively depletes in the dry phase in response to these human actions.

4.3. 2D VS Models

Figure 5 shows the 2D VS models obtained in sites 1 and 3 (Figure 1d) on 23 June 2014 (Table 2). Prospecting depth was 30 m in MASW1 (Figure 5a) and 27 m in MASW3 (Figure 5b). There is not MASW2 survey in site 2 because the operation was cancelled due to a heavy storm occurred while acquiring data, thus provoking a battery discharge that left the equipment inoperable. Both MASW1 and MASW3 were interpreted recurring to (i) reference VS values compiled from the scientific literature for equivalent lithologies [34,59] as in Table 4, (ii) local geological information after [43,46–48], and (iii) detailed geological data for Pliocene to Holocene formations deduced from geotechnical soundings S1 to S3 [40] as in Figure 1d.

In both MAWS1 and MASW3, VS is in the 50–950 m s^{-1} range. In general, VS increases in depth according to the increasing age and compaction of sediments, from less than 200 m s^{-1} for Holocene, 200–500 m s^{-1} for Pleistocene, to more than 500 m s^{-1} for Pliocene formations. This vertical VS distribution correlates well with regional geological information reported by [47,48] and geotechnical data from soundings S1 to S3 [40]. The horizontal continuity of this vertical VS distribution is frequently interrupted due to sedimentary processes (e.g., lateral facies changes, erosive channels) and action of faults, which determine the accommodation space for Holocene sedimentation, as discussed in Section 5.2.

In MASW1, the uppermost 8 m (northern and southern valley boundaries) and 10 m (central valley) thick VS < 200 m s^{-1} values are attributed to aquifer H, the underlying 18 m (central valley) and 20 m (valley boundaries) thick 200 < VS < 500 m s^{-1} to Pleistocene sands and gravels, and the deeper 5 m (central valley) and 10 m (valley boundaries) thick VS > 500 m s^{-1} to Pliocene formations (Figure 5a).

In MASW3, VS data acquisition at distances 131–220 m was interrupted due to the impossibility of the equipment to work in the inundated Cascalheira stream bed (Figure 5b). This VS model gap does not compromise the geological interpretation because the ERT3 survey partially covered that sector, as described in next Section 4.4. In MASW3, the uppermost 9 m (northern and southern valley boundaries) and 12 m (central valley) thick VS < 200 m s^{-1} values are attributed to aquifer H, the underlying 20 m (central valley) and 15 m (northern valley boundary) thick 200 < VS < 500 m s^{-1} to Pleistocene sediments, and the deeper 7 m (northern valley boundary) thick VS > 500 m s^{-1} to Pliocene formations (Figure 5b). At distances 220–310 m, the Pleistocene–Pliocene boundary is not identified because it is below the prospecting depth. As in site 1, site 3 shows contrasted VS values associated to geological formations having different age and compaction. However, at the southern sector, Holocene sediments and underlying Pleistocene sand dunes show similar VS < 200 m s^{-1}, thus limiting to identify its boundary. As described in next Section 4.4, ERT enables disambiguating the boundary of these formations.

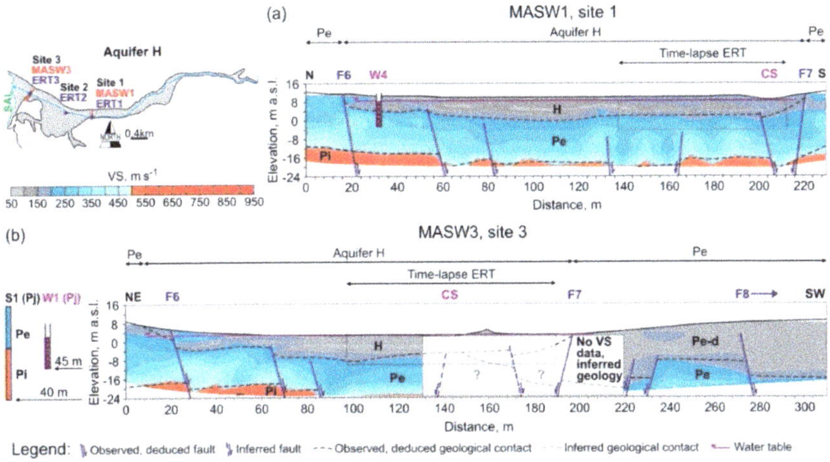

Figure 5. MASW1 and MASW3 surveys at sites 1 (**a**) and 3 (**b**), respectively. A preliminary geological interpretation of VS values is included, showing projected (Pj) log of geotechnical sounding S1 after [40], and water table recorded in September 2014 in wells W4 (site 1) and W1 (site 3) as in Figure 4. Profiles are topographically corrected and its vertical-to-horizontal scale ratio is 1:1. CS, Cascalheira Stream; H, Holocene; Pe, Pleistocene; Pe-d, Pleistocene sand dunes; Pi, Pliocene; and F1 to F3, NW–SE strike-slip faults. The dotted-line rectangle is the area covered by time-lapse ERT surveys at sites 1 and 3.

The VS models statistics are in Table 5. Average VS lightly decreases from MASW1 to MASW3 as the influence of high VS values from Pliocene materials decreases because they were below its prospecting depth. The coefficient of variation of VS shows similar natural variability of VS in both MASW1 and MASW3 associated to a similar, predictable vertical geological structure. As deduced, VS is predictable enough to produce a confident near-surface geological definition.

Table 5. VS models statistics.

Profile ID [1]	AV VS [2]	SD VS [3]	CV VS
MASW1	273.1	161.4	0.59
MASW3	215.1	126.8	0.59

[1] ID and location as in Figure 1d. [2] AV and SD are average and standard deviation of VS in m s^{-1}. [3] CV is dimensionless coefficient of variation of VS (SD-to-AV ratio) as a fraction.

4.4. 2D ER Models

Sequential 2D ER models are used to refine geometry of aquifer H. As described in Section 4.2, when porous geological media hardly contribute, ER changes can be attributed to transient GER changes produced by temporal water transferences and fluxes [29,31]. The ER models showed in Figure 6 were interpreted recurring to (i) previous VS models (Figure 5) and geological data [40,43,47,48] to infer geology, and (ii) natural processes and some human actions modifying GER to deduce transient water transferences and fluxes (Figure 4).

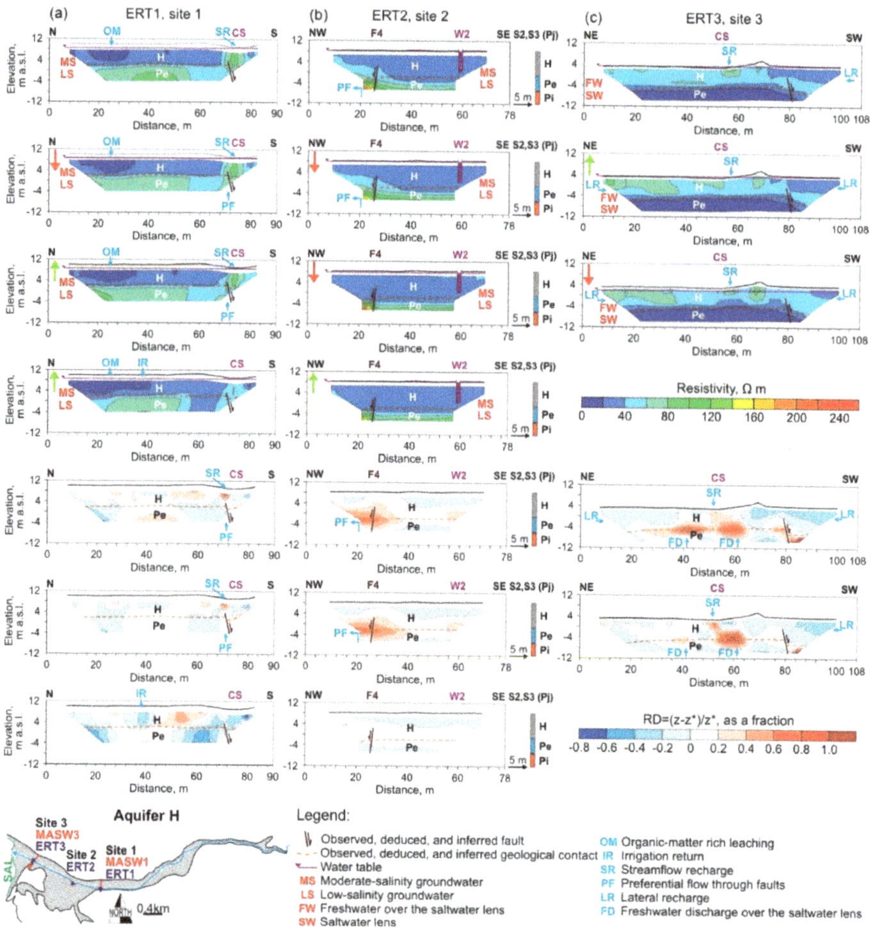

Figure 6. At sites 1 (**a**), 2 (**b**), and 3 (**c**), time-lapse ERT1, ERT2, and ERT3 surveys in March, June, September, and December 2014 (rows 1 to 4), and relative difference, $RD = (z - z^*)/z^*$, of nodal ER data from March, June, and December lapses (z) regarding nodal ER data from the September lapse (z^*) (rows 5 to 7). The projected (Pj) logs of geotechnical soundings S2 and S3 after [40], and water table recorded in each lapse in sites 1 (well W4), 2 (well W2), and 3 (well W1) as in Figure 4 are shown; in the top left-hand corner of plots, green and red vertical arrows show, respectively, water-table raise and depletion relative to previous lapse. Preliminary geological and hydrogeological interpretations of ER models after integration of geological findings from VS models (Figure 5) and aquifer conceptualization (Figure 2) are included; red acronyms denotes groundwater types, and blue arrows and acronyms denote natural processes and human-induced actions determining transient water transferences and fluxes. Profiles are topographically corrected and its vertical-to-horizontal scale ratio is 1:1. CS, Cascalheira Stream; H, Holocene; Pe, Pleistocene; and F8, NE–SW strike-slip fault.

In sites 1 to 3 (Figure 1d), time-lapse ERT were in March, June, September, and December 2014 (Table 1) attending to the minimum frequency for optimal ERT surveying calculated in Section 4.1. For reliable comparisons in each site, the 2D ER models keep the same prospecting distance and depth (same grid). Prospecting depth was 13 m in ERT1 (Figure 6a), 15 m in ERT2 (Figure 6b), and 13 m in ERT3 (Figure 6c). In site 1, the December 2014 survey was cancelled because the SAL overflowed and inundated the ERT space after heavy rainfall events while the coastal sandbar was closed.

For all time lapses in ERT1 to ERT3 surveys, the ER dataset is positively skewed in the 1.6–237.9 Ω m range; the average value being 41.03 Ω m (Figure 6). This experimental ER range agrees with the theoretical GER baseline deduced in Section 4.2. Thus, ER datasets in sites 1 to 3 are hardly influenced by subsurface geology. In general, ER is 32–49 Ω m in ERT1, 21–44 Ω m in ERT2, and 37–38 Ω m in ERT3. The 2D ER models identify well (i) the vertical distribution of Holocene and Pleistocene formations deduced in Figure 5 and its horizontal interruptions mostly due to faults; and (ii) the general groundwater types, both low-salinity Pleistocene and moderate salinity Holocene in sites 1 and 2 and saltwater Pleistocene and freshwater Holocene in site 3 as conceptualized in Figure 2. In site 3, sequential time-lapse ERT3 enabled to (i) infer geology due to interruption in VS acquisition at distances 130–220 m in MASW3, and (ii) disambiguate aquifer H from Pleistocene sand dunes having similar VS < 200 m s^{-1} (Figure 5b).

In detail, most of ER values in aquifer H are 20–40 Ω m in ERT1 (inland site) and in ERT2, and 40–60 Ω m in ERT3 (coastal site). This ER distribution apparently contradicts the expected decreasing inland gradient of GEC inferred by the decreasing inland gradient of atmospheric bulk deposition EC reported in the southwest coast of Portugal [39]. Thus, another process such as low-salinity (high-resistivity) groundwater transference from the hydraulically connected underlying Pleistocene formation through faults is proposed to explain this ER pattern, as illustrated in ERT2 (Figure 6b). Most of ER values in Pleistocene formations are 50–70 Ω m in ERT1 (inland site) and in ERT2, and 1–20 Ω m in ERT3 (coastal site) (Figure 6). In site 3, the lowest ER values in Pleistocene formation are attributed to the saltwater lens conceptualized in Figure 2, whereas ER values in the 20–40 Ω m range are attributed to a thin freshwater-saltwater interface (Figure 6c).

As above described, ER changes inside steady ER shapes are associated to transient GER changes due to water transferences and fluxes varying over time, both including natural processes such as stream flow recharge, preferential flow through faults, lateral recharge, and freshwater discharge over the saltwater lens, as well as human-induced actions such as irrigation return and organic matter-rich leaching (Figure 6). In order to quantify how and where natural processes and human-induced actions modify ER over time, in each site, nodal ER data from March, June, and December lapses (z) were compared to nodal ER data from the September lapse (z*). The relative difference is:

$$RD = (z - z^*)/z^*. \quad (3)$$

RD is mapped in Figure 6. The rationale for reference lapse choosing was that geological formations must have the lowest GEC mass flow and storage in order to minimize the influence of groundwater transference among them. September 2014 was the selected reference lapse for comparisons (Figure 4c). RD values enable to show both natural processes and human-induced actions modifying ER.

Statistics of time-lapse ER models and RD values are in Table 6. Average ER follows the same evolution in sites 1 and 2, with higher values in March and June and lower in December regarding those observed in September. In site 3, lower ER values in March are attributed to the delayed influence of brackish groundwater discharge stored after the SAL opened, whereas in December 2014 (this site cannot be surveyed), the same evolution to sites 1 and 2 is presumed. It is important to note that ER models (Table 6) are delayed regarding atmospheric bulk deposition EC values (Figure 4b). This fact corroborates that average groundwater turnover time is higher than three months, long after the dry phase (December 2014), because the ER evidences of current aquifer recharge and groundwater transference from underlying geological formations have not yet arrived to aquifer H. The coefficient of variation of ER evidences this temporal predictability.

Table 6. Time-lapse ER models statistics.

Profile ID [1]	Time-Lapse [2]	AV ER [3]	SD ER [3]	CV ER [4]	AV EC [3]	RD ER [5]
ERT1	March	44.06	23.06	0.52	300	0.043
	June	43.37	22.78	0.53	310	0.028
	September	42.15	22.83	0.54	320	0
	December	38.18	19.96	0.52	340	−0.104
ERT2	March	48.85	33.18	0.68	250	0.034
	June	49.00	31.51	0.64	250	0.037
	September	47.17	37.12	0.79	270	0
	December	45.69	30.78	0.67	270	−0.032
ERT3	March	37.34	20.65	0.55	580	−0.027
	June	38.40	22.69	0.59	580	0.002
	September	38.33	22.87	0.60	590	0

[1] ID and location as in Figure 1d. [2] Dates are referred to year 2014 as in Table 2. [3] AV and SD are average and standard deviation of ER in Ω m and of EC in μS cm^{-1}. [4] CV is dimensionless coefficient of variation of ER (SD-to-AV ratio) as a fraction. [5] $RD = (z - z^*)/z^*$ is dimensionless relative difference of nodal ER data in a time-lapse ERT (z) regarding nodal ER data in the September 2014 reference time-lapse ERT (z^*), as a fraction.

5. Discussion

5.1. Performance of VS and ER Models

Performance analysis is crucial in measuring the quality of prognostics in different modeling fields and consists of different statistical calculations between measured (M) and predicted (P) data. In this paper, performance analysis allowed us to evaluate the prediction ability of the VS and ER models, providing comparable quotas of uncertainty, which could not be done directly using the errors provided by the geophysical software. In geosciences, there are no defined protocols or consensus on which statistics or group of statistics are the best to evaluate performance of models [65–67]. For variables VS and ER, multiple expressions for determining point distance between M and P data and comparable quotas of the relative error between models are used to produce a complete evaluation of models performance and error distribution, as proposed by [66–68].

Reliability of the performance and error expressions was analyzed, provided M and P data sets were normal or lognormal distributed [6,52,67]. The Kolmogorov–Smirnov goodness-of-fit was used for testing the data sets normality. The M and P data sets showed quite skewed distributions (rows 1 and 2 in Figure 7). Their logarithm proved to fit close-to-normal distributions (rows 3 and 4 in Figure 7). The logarithm M and P data pairs were plotted and the results proved to fit close to the theoretical linear 1:1 ratio (row 5 in Figure 7). After this analysis, normal M and P data sets were lognormal converted, the performance statistics and error expressions were applied to these lognormal data sets, and results were reverted to the original magnitude of variables.

For VS and ER models efficiency criteria, Nash–Sutcliffe efficiency coefficient (NSE), logarithmic form of NSE (lnNSE), coefficient of determination (R^2), percent bias (PBIAS), root-mean-square error (RMSE), RMSE relative to standard deviation of the measured data (RSR), mean absolute error (MAE), and mean relative error (MRE) were used (Table 7). Description of statistics is detailed in [66,68–70].

Normalizing the statistics facilitates the comparison of VS and ER models having different exploration scales, e.g., MASW and ERT surveys having equal spacing grid but different number of nodes determining different exploration length and depth. Though there is no consistent means of normalization in the literature, common choices are the mean, the range (the difference in maximum and minimum values), and the interquartile range (the difference in 0.75 and 0.25 quartiles) of the M data set. Here, the mean was selected as quotient despite that the interquartile range is less responsive to outliers in [68]. The term coefficient of variation (CV) refers to a normalization using the mean value, and may be used to avoid ambiguity regarding other normalizing procedures. This is analogous to the coefficient of variation with a given absolute error statistics, taking the place of the standard deviation of the measured data (STD), which is the most used statistics for normalizing errors [67]. CV facilitates comparable quotas of the relative error between models having different number of data. For this purpose, CV must be additionally corrected by applying the factor n/n', where n is the total number of data of a given model and n' is the total number of data of the biggest explored model: MASW3 for VS models (Figure 5) and ERT3 for ER models (Figure 6); n values are in Figure 7 and in Table 7. Normalized MAE (CVMAE), RMSE (CVRMSE), and STD (CVSTD) were used (Table 7).

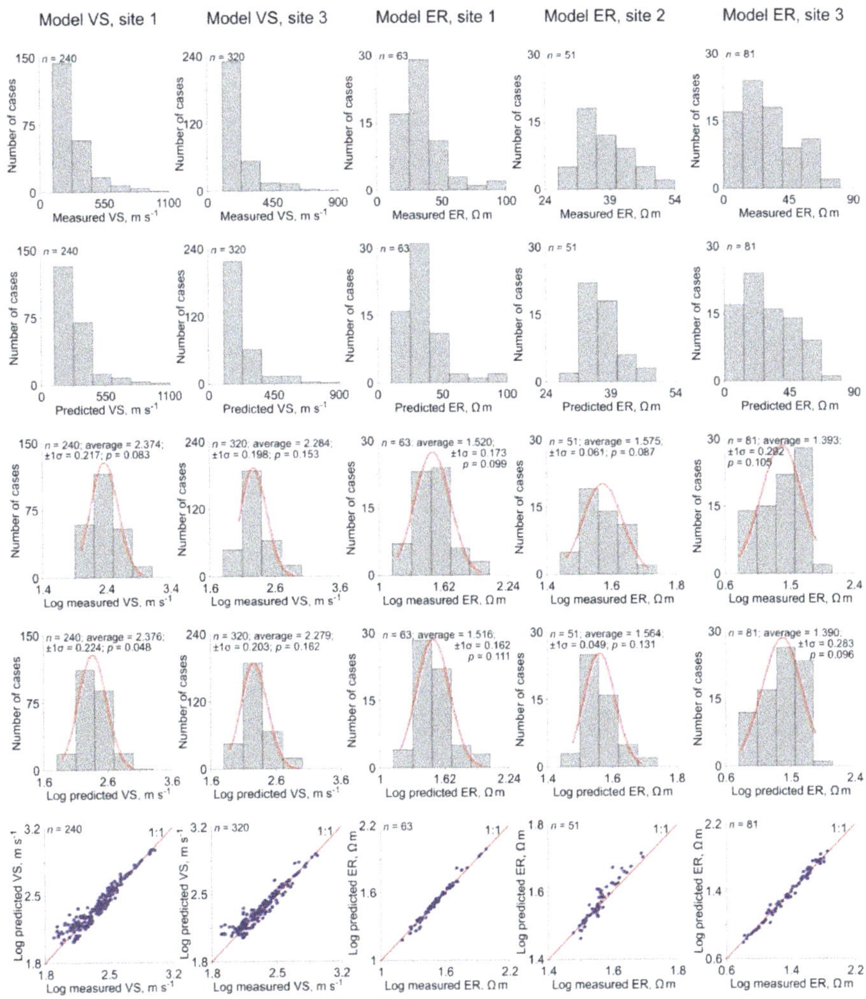

Figure 7. For VS and ER models in sites 1 to 3 (columns 1 to 5), histograms of measured (M) and predicted (P) data sets (rows 1 and 2), histograms of logarithmic M and P data sets with the fitted lognormal density functions (rows 3 and 4), and logarithm M vs. P data pairs with the 1:1 relationship. n = number of data. $\pm 1\sigma$ = standard deviation. p = p-value from a Kolmogorov–Smirnov goodness-of-fit test.

Table 7. Equations, ranges, optimal values, and results for VS and ER models performance statistics and normalized errors at sites 1 to 3, after [66,68–70]. For ER, the time-lapse ERT from September 2014 was selected.

Statistics and Equation [1]	Definition, Range, and Match	Site 1		Site 2	Site 3	
		VS	ER	ER	VS	ER
NSE: Nash–Sutcliffe efficiency coefficient $= 1 - \frac{\sum_{i=1}^{n}\left(D_{m\,i}-D_{p\,i}\right)^2}{\sum_{i=1}^{n}\left(D_{m\,i}-\overline{D_m}\right)^2}$	NSE indicates a perfect match between measured (M) and predicted (P) data. NSE ranges from $-\infty$ to 1. Match is satisfactory from >0.7.	0.90	0.98	0.84	0.90	0.98
lnNSE: logarithmic form of NSE $= 1 - \frac{\sum_{i=1}^{n}\left(\ln(D_{m\,i})-\ln(D_{p\,i})\right)^2}{\sum_{i=1}^{n}\left(\ln(D_{m\,i})-\ln(\overline{D_m})\right)^2}$	lnNSE emphasizes low values, and NSE the high ones. Match is satisfactory from >0.7.	0.88	0.97	0.84	0.88	0.98
R^2: coefficient of determination $= \left(\frac{\sum_{i=1}^{n}\left(D_{m\,i}-\overline{D_m}\right)\left(D_{p\,i}-\overline{D_p}\right)}{\sqrt{\sum_{i=1}^{n}\left(D_{m\,i}-\overline{D_m}\right)^2}\sqrt{\sum_{i=1}^{n}\left(D_{p\,i}-\overline{D_p}\right)^2}}\right)^2$	R^2 indicates the degree of linear relationship between M and P data. R^2 ranges from 0 to 1. Match is satisfactory from >0.7.	0.90	0.98	0.90	0.91	0.98
PBIAS: percent bias $= \frac{\sum_{i=1}^{n}\left(D_{m\,i}-D_{p\,i}\right)}{\sum_{i=1}^{n}\left(D_{m\,i}\right)} \times 100$	PBIAS calculates the average tendency of the P data to be higher or lower than their M counterparts. The optimal value is 0. Perfect match is 0. Acceptable match is in the ±25% range.	−0.07	0.24	0.73	0.24	0.20
RMSE: root-mean-square error $= \sqrt{\sum_{i=1}^{n}\left(D_{m\,i}-D_{p\,i}\right)^2}$	RMSE calculates the precision of the P data. Perfect match is 0. Increasing RMSE values indicate that matching worse, typically due to outliers.	11.99	1.61	1.49	12.89	2.14
RSR: RMSE relative to standard deviation of the measured data $= \frac{RMSE}{STD_m} = \frac{\sqrt{\sum_{i=1}^{n}\left(D_{m\,i}-D_{p\,i}\right)^2}}{\sqrt{\sum_{i=1}^{n}\left(D_{m\,i}-\overline{D_p}\right)^2}}$	RSR ranges from 0 to ∞. The lower the RSR, the lower the RMSE and the better the model performance. Acceptable match is <0.5.	0.01	0.07	0.54	0.00	0.01
MAE: mean absolute error $= \frac{1}{n}\sum_{i=1}^{n}\left\|D_{p\,i}-D_{m\,i}\right\|$	MAE is the absolute difference in the P and M data. Perfect match is 0.	1.13	1.04	1.04	1.11	1.07
MRE: mean relative error $= \frac{1}{n}\sum_{i=1}^{n}\frac{\left\|D_{p\,i}-D_{m\,i}\right\|}{D_{m\,i}}$	MRE is the relative difference in the P and M data. Perfect match is 0.	0.02	0.01	0.01	0.02	0.02

Table 7. *Cont.*

Statistics and Equation [1]	Definition, Range, and Match	Site 1		Site 2	Site 3	
		VS	ER	ER	VS	ER
CVMAE: normalized MAE $= \frac{MAE}{D_m}$	Perfect match is 0. Acceptable match is <0.3.	0.01	0.04	0.04	0.01	0.04
CVRMSE: normalized RMSE $= \frac{RMSE}{D_m}$	Perfect match is 0. Acceptable match is <0.3.	0.07	0.06	0.06	0.07	0.09
CVSTD: normalized STD $= \frac{STD_m}{D_m}$	Perfect match is 0. Acceptable match is <0.3.	0.01	0.06	0.05	0.01	0.08
n		240	63	51	320	81
MIN_m		102.00	15.28	29.04	113	6.33
MIN_p		79.25	15.85	30.12	80.01	75.90
MAX_m		948.00	98.17	51.93	831.00	6.48
MAX_p		1015.61	96.88	49.48	867.02	72.24
$\overline{D_m}$		236.67	33.11	37.61	192.46	24.73
$\overline{D_p}$		237.60	32.82	36.64	190.03	24.57

[1] Over the exploration space (a given geophysical profile), n is total number of data, MIN_m and MAX_m are minimum and maximum measured (M) values, MIN_p and MAX_p are minimum and maximum predicted (P) values, and $\overline{D_m}$ and $\overline{D_p}$ are mean of the M and P data sets. $D_{m\,i}$ and $D_{p\,i}$ are M and P value at observation point i. RMSE, MAE, MIN_m, MIN_p, MAX_m, MAX_p, $\overline{D_m}$, and $\overline{D_p}$ are in Ω m for ER models, and in m s^{-1} for VS models. NSE, lnNSE, R^2, RSR, and MRE are dimensionless. PBIAS, CVMAE, CVRMSE, and CVSTD are dimensionless, as a fraction.

Performance of VS and ER models for geological model conceptualization, aimed at supporting groundwater flow modeling, departs from our initial ability to identify and segregate the influence of possible outliers caused by natural heterogeneity of variables VS and ER. As described in Section 4.2, VS is site-specific steady, whereas ER is site-specific transient, the latter varying over time and requiring periodic monitoring. For this reason, frequency of time-lapse ERT surveying and groundwater monitoring was adapted to the particular groundwater quantity (flow) and quality (EC) temporal dynamics of aquifer H; see Section 4.1. The evaluation of this optimal surveying frequency is crucial for the production of accurate geological models. Performance statistics of VS and ER models show satisfactory or very satisfactory matches between M and P data sets (Table 7), i.e., P data are less biased than their M counterparts, the relative difference being in the positive 0.01–0.02 range. Performance of the VS models is somewhat greater than of the ER models (Table 7).

This performance analysis provides two important remarks: (1) theoretical fitting functions do not introduce noticeable mismatching between M and P data, i.e., do not generate spurious values, because the number and magnitude of original outliers is low to negligible; and (2) computation and fitting processes reproduce well the homogeneity of this alluvial medium, in which a general predictable behavior of variables VS and ER in non-surveyed sites can be anticipated from the results obtained in the surveyed ones, as show in next Section 5.2. Performance of the VS and ER models was better to that reported in similar experiences using the same methodology for MASW [59,63,64] and ERT [71–73]. As deduced, the VS and ER models are reliable enough to produce a confident geological model conceptualization for groundwater modeling purposes.

As introduced in Section 1, errors in data acquisition for aquifer conceptualization are rarely segregated from the overall uncertainty appraisal of groundwater modeling tools. Aquifer conceptualization includes three stages subjected to error: (1) geometry definition, (2) acquisition of hydraulic data, and (3) evaluation of water balance components [15–17]. These errors combine to generate a background error from which the groundwater model simulations add uncertainty associated to the inherent natural variability of environmental variables, the possible smoothing and bias introduced using fitting functions, and the mapping errors due to adopted spatial functions. In the case of aquifer geometry definition, the use of geophysical techniques does not cancel nor reduce its input error, but it generates a numerical data base (M and P data sets) that enables to deduce the error produced during data inversion. This error can be measured when M and P data are compared, and becomes an advantage comparatively to the sole use of raw geological data and qualitative, large-scale geological mapping for the purpose of segregating the error of aquifer geometry.

Normalized errors (CVMAE, CVRMSE, CVSTD) of VS and ER models are in the 0.01–0.09 range (Table 7). In particular, CVSTD is 0.01 for VS models and varies in the 0.05–0.08 range for ER models. These magnitudes are lower than those that can be deduced from the VS data sets reported in similar MASW experiences [59,63,64] and the normalized errors of ER models in diverse ERT surveys [68–70] experiences. These values can be used as manageable quotas of the error that can be segregated from overall groundwater models uncertainty without substantially affecting its simulations accuracy.

5.2. The Geological Model of the Cascalheira Stream Alluvial Aquifer

MASW and ERT geophysical techniques were integrated to define the geological model of aquifer H. The VS models were intended to this purpose exclusively, whereas the ER models were used to disambiguate geological structures having similar VS and different ER, and to complete geological information in areas not covered by MASW.

As described in Section 2.3, local faulting narrowly reproduces the regional structural setting determined by the conjugate NW–SE and NE–SW strike-slip fault systems (Figure 8a) [44,45]. These fault systems affect the Upper Miocene—Quaternary sedimentary record, thus determining the accommodation space for Holocene sedimentation in valleys and plains in the southwest coast of Portugal [43,47,48]. As shown in VS (Figure 5) and ER (Figure 6) models, the NW–SE strike-slip fault system generates small horsts and grabens perpendicular to the coast (Figure 8b). From north to south, the VS and ER models identified three small NW–SE faults called F6, F7, and F8; F8 was also inferred from regional geological mapping [43,47,48]. Out the area covered by MASW and ERT surveys, another small NW–SE fault called F9 was inferred after regional geological mapping and direct field observation. F6 and F8 are SW-vergent, whereas F7 and F9 are NE-vergent. The NE–SW strike-slip fault system is NW-vergent, is younger than the NW–SE system, and determines the progressive deepening of the geological formations toward the coast (Figure 8b). From East to West, two small NE–SW faults called F1 and F2 were inferred from regional geological mapping [43,47,48], two others called F3 and F5 were deduced after direct field observation, and a fifth called F4 was inferred after analyzing time-lapse ERT at site 2. The conjugation of NW–SE and NE–SW fault systems compartmentalizes the area into small NW-vergent blocks. The Cascalheira Stream valley is the result of these conjugate strike-slip fault systems; first, the NW–SE system generates a small graben, and later, the NE–SW system deepens and widens the aquifer H toward the coast. This structural scheme is described in Figure 8b.

In detail, the NW–SE faults F6 and F7 at site 1 (MASW1) were identified (Figure 8c). At site 3, the traces of NW–SE faults F6, F7, and F8 (MASW3) were also identified (Figure 8c). At sites 1 and 3, no NE–SW fault was identified because the MASW and ERT surveys were almost parallel to these faults (Figure 8c). However, at site 2, the ER model enabled us to deduce the NE–SW fault F4 as a preferential path for low-salinity water transference from Pleistocene formations to aquifer H (Figure 6b). At site 3, the ER model enabled us to infer geology after VS model gap in MASW3 at distances 130–220 m and to disambiguate the geological structure of this profile sector, which is infilled by Holocene sediments and Pleistocene sand dunes having similar VS (MASW3, Figure 5b) but different ER (ERT3, Figure 6c). After the ER models, some sedimentological features of hydrological interest have been deduced. At site 1, the current stream channel appears migrated towards the southern valley boundary, probably as the central valley was progressively infilled since early Holocene. This feature is also observed at site 3, determining that freshwater discharge from aquifer H over the saltwater lens in Pleistocene formations flow through what seems to be two incisive paleochannels (Figure 6c).

Figure 8c presents the geological model of aquifer H. As shown, the Cascalheira Stream valley is encased into a small graben structure contoured by small NW–SE and NE–SW strike-slip fault systems. In detail, aquifer H occupies one small graben bounded by faults F6 and F7 upstream and two small grabens (bounded by faults F6–F7 and F8–F9) separated by a small horst (bounded by faults F7–F8) downstream. The increasing accommodation space for Holocene sedimentation is attributed to the distensive motion of (i) faults F6 to F9 which progressively deepen the Pliocene and Pleistocene materials enclosed into the two small grabens, and (ii) faults F1 to F5 which progressively deepen the geological structures toward the coast. The consequence of this faulting is a typical estuarine morphology infilled by Upper Miocene to Quaternary sediments whose thickness and width increase downstream. The aquifer H thickness passes from 3–10 m at site 1, 5–10 m at site 2, to 8–12 m at site 3, with some VS increase (Figure 5) and ER decrease (Figure 6) in this bearing as expected in a coastal alluvial where clay-rich materials (higher VS) can be dragged bigger distances than coarse ones and salinity of marine aerosol decreases inland (higher GER), respectively. Width passes from 140 m at site 1 to 240 m at site 3.

Despite that, the described NW–SE and NE–SW faults are too small to be included into the Official Geological Mapping of Portugal at scale 1:50,000 [43]—only the NW–SE F6 and F9 and the NE–SW F1 and F2 appear in this official mapping—its strike-slip motions are long enough to produce the accommodation space for aquifer H (Figure 8c). This unconfined alluvial aquifer sustains the SAL, a protected GDE space.

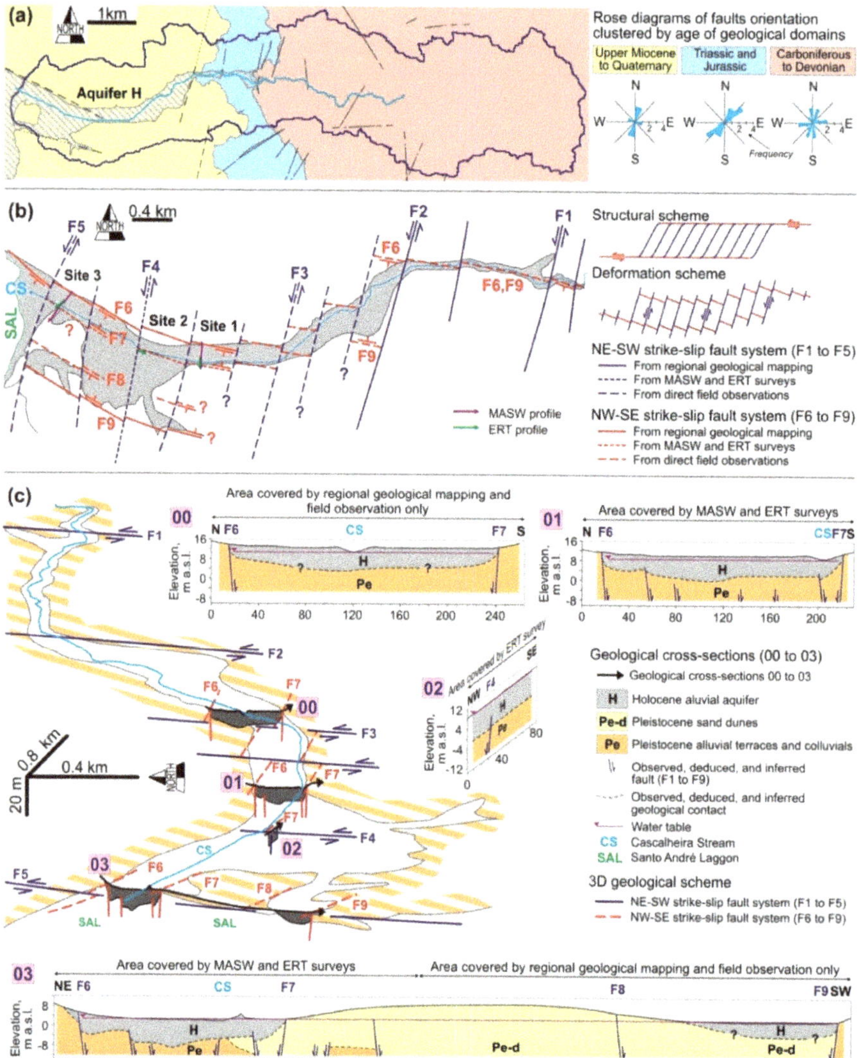

Figure 8. (**a**) Regional tectonic setting of CSB area, after [43–45], showing rose diagrams of faults orientation clustered by age of geological domains as in Figure 1b. (**b**) Local tectonic setting of aquifer H after [40,43,47], direct field observations, aerial photographs, and MASW and ERT surveys, showing theoretical structural and deformation schemes, and the conjugate NW–SE and NE–SW strike-slip fault systems. (**c**) Geological model of aquifer H, showing geological cross-sections 01 (site 1), 02 (site 2), and 03 (site 3) from integrated MASW (Figure 5) and time-lapse ERT (Figure 6) surveys at sites 1 to 3 and direct field observations, and a new geological cross-section called 00 and the southern end of the geological cross-section 03 inferred from geological mapping and direct field observation only; the vertical-to-horizontal scale ratio being 1:2.

6. Conclusions

Integrated MASW and time-lapse ERT geophysical techniques enabled building a predictive geological model and deducing some transient groundwater features of the CSB Holocene alluvial aquifer (aquifer H) in the southwest coast of Portugal. Findings from 2D VS and ER models were completed with official regional geological information, geological and geophysical data compiled from the scientific literature, and direct field observations. The conjugate NW–SE and NE–SW strike-slip fault systems determine compartmentalization of underlying geological structures and the subsequent accommodation space for Holocene sedimentation. The NW–SE system deepens the geological structures toward the coast, whereas the NE–SW system generates small horsts and grabens, the aquifer H being encased into one of these small grabens. From upstream to downstream, aquifer H thickness and width increase from 10 m to 12 m and from 140 m to 240 m, respectively.

The two 2D VS models in sites 1 (upstream) and 3 (downstream) were addressed at defining the near-surface geological structure of aquifer H and underlying Pleistocene and Pliocene formations. Since VS is a site-specific steady variable which does not depend on ambient changes, the VS models can be used to interpret geological structures without the groundwater component influence. In interpreting VS models, some reference VS values compiled from scientific literature for equivalent lithologies were used. The ER models were aimed at inferring some shallowest geological structures after some VS gaps and disambiguating geological structures having different age, similar VS, and different ER. Unlike VS, ER is a site-specific variable which is subjected to ambient changes. In coastal porous media, ER is primarily governed by GER, which in turn depends on the recharge water salinity resulting from the variable combination of atmospheric bulk deposition salinity and actual evapotranspiration rates, which are predictable variables over space and time. These predictive variations allowed us to disambiguate inland recharged Holocene alluvial from locally recharged coastal Pleistocene sand dunes at the aquifer H outlet. For proper disambiguating, the time-lapse ERT from September 2014, in which GEC mass flow and aquifer storage variations are minimal, was selected.

Groundwater predictions from numerical models can be quite inaccurate when inherent uncertainty of data used for aquifer conceptualization is neither appraised nor considered. In contrast to the sole use of qualitative raw geological data and mapping to prepare a geological model, normalized errors of the VS and ER models mean a quantitative quota of uncertainty from which groundwater models will add other types of uncertainty. The performance analysis of the VS and ER models shows satisfactory to very satisfactory matches between measured and predicted data. This means that the Pliocene to Holocene geological structures were well-conceptualized and they are homogeneous enough to predict the magnitude of VR and ER in non-surveyed areas from the results obtained in the surveyed ones. Normalized errors of the VR and ER models are in the 0.01–0.09 range, which means a manageable quota of error that can be segregated from overall groundwater models uncertainty without substantially affecting its simulations accuracy. The geological model of aquifer H is aimed at supporting the CSB—SAL groundwater numerical model, which is in progress for ecological purposes. This multidisciplinary methodology seeks to improve the design of shallow groundwater research in GDE preservation policies.

Author Contributions: Conceptualization, M.C.P., F.J.A. and L.R.; Methodology, M.C.P., F.J.A., P.M.-P, J.P.-C.; Validation, M.C.P and F.J.A.; Formal Analysis, M.C.P., F.J.A. and A.M.; Investigation, M.C.P., F.J.A., A.M., P.M.-P. and J.P.-C.; Resources, M.C.P., F.J.A., P.M.-P. and L.R.; Data Curation, M.C.P. and J.P.-C.; Writing-Original Draft Preparation, M.C.P. and F.J.A; Writing-Review & Editing, M.C.P., F.J.A. and P.M.-P.; Visualization, M.C.P and F.J.A.; Supervision, F.J.A. and L.R.; Project Administration, L.R.; Funding Acquisition, M.C.P., F.J.A. and L.R. All authors have read and agreed to the published version of the manuscript.

Funding: This research was funded by the Portuguese FCT research projects PTDC/CLI/72585/2006 (Climarid) and PTDC/AAC-AMB/104639/2008 (GroundScene). MCP was funded by the Portuguese FCT grant number SFRH/BD/75327/2010.

Acknowledgments: This work is dedicated to the memory of Luís Ribeiro, whose devotion to groundwater research leaves a legacy of multidisciplinary knowledge to the scientific community and to the public in general. Research funded by the Portuguese FCT research projects PTDC/CLI/72585/2006 (Climarid) and PTDC/AAC-AMB/104639/2008 (GroundScene). M.C.P. is grateful to the Portuguese FCT for the PhD Grant SFRH/BD/75327/2010. The authors are grateful to Eng. Vítor Paz and Carlos Costa for assisting ERT field surveying and monitoring, to Mohammad Farzamian from IDL-FCUL for advising time-lapse ERT data inversion, and to SAL authorities for providing access to protected areas.

Conflicts of Interest: The authors declare no conflict of interest.

References

1. Sabater, S.; Barceló, D. *Water Scarcity in the Mediterranean. Perspectives under Global Change*; Springer: Berlin/Heidelberg, Germany, 2010; p. 234. [CrossRef]
2. Custodio, E. Coastal aquifers of Europe: An overview. *Hydrogeol. J.* **2010**, *18*, 269–280. [CrossRef]
3. Martínez-Valderrama, J.; Ibáñez, J.; Alcalá, F.J. AQUACOAST: A simulation tool to explore coastal groundwater and irrigation farming interactions. *Sci. Programming-Neth* **2020**, *2020*, 092829. [CrossRef]
4. Dalin, C.; Wada, Y.; Kastner, T.; Puma, M.J. Groundwater depletion embedded in international food trade. *Nature* **2017**, *543*, 700–704. [CrossRef] [PubMed]
5. Alcalá, F.J.; Martínez-Valderrama, J.; Robles-Marín, P.; Guerrera, F.; Martín-Martín, M.; Raffaelli, G.; Tejera de León, J.; Asebriy, L. A hydrological-economic model for sustainable groundwater use in sparse-data drylands: Application to the Amtoudi Oasis in southern Morocco, northern Sahara. *Sci. Total Environ.* **2015**, *537*, 309–322. [CrossRef]
6. Alcalá, F.J.; Martín-Martín, M.; Guerrera, F.; Martínez-Valderrama, J.; Robles-Marín, P. A feasible methodology for groundwater resource modelling for sustainable use in sparse-data drylands: Application to the Amtoudi Oasis in the northern Sahara. *Sci. Total Environ.* **2018**, *630*, 1246–1257. [CrossRef]
7. Salvador, N.; Costa, L.; Hugman, R.; Monteiro, J.P.; Stigter, T.; Nunes, L.; Duarte, D. Monitoring and modelling groundwater contributions to dependent ecosystems—The case study of the Santo André coastal lagoon. In *Mudança de Planos, Proceedings of the VIII Congresso Ibérico sobre Planeamento e Gestão da Água, Lisboa, Portugal, 5–7 December 2013*; Universidade Lusíada Editora: Lisboa, Portugal, 2013; Available online: http://revistas.lis.ulusiada.pt/index.php/8cigpa/article/view/386/pdf_65 (accessed on 1 May 2020).
8. Paz, C.; Alcalá, F.J.; Carvalho, J.M.; Ribeiro, L. Current uses of ground penetrating radar in groundwater-dependent ecosystems research. *Sci. Total Environ.* **2017**, *595*, 868–885. [CrossRef]
9. WFD. Directive 2000/60/EC of the European Parliament and of the Council of 23 October 2000 establishing a framework for Community action in the field of water policy. *Off. J. Eur. Communities* **2000**, *327*, 1–73.
10. Cancela da Fonseca, L.; Chainho, P.; Félix, P.M.; Correia, M.J.; Costa, J.L.; Chaves, M.L.; Domingos, I.; Lopes, V.; Mirra, C.; Castro, J.; et al. *Modelação de Cenários de Exploração em Aquíferos de Zonas Costeiras: Efeitos na Biodiversidade de Lagoas e Respectivas Ribeiras Como Ecossistemas Dependentes de água Subterrânea—Groundscene–PTDC/AAC-AMB/104639/2008–Relatório Final de Execução Científica do Projecto*; Fundação para a Ciência e Tecnologia: Lisboa, Portugal, November 2013; p. 62.
11. Chainho, P.; Félix, P.M.; Correia, M.J.; Fernandes, C.M.; Costa, J.L.; Chaves, M.L.; Stigter, T.; Hugman, R.; Salvador, N.; Costa, L.; et al. Projecto Groundscene: Biodiversidade de lagoas costeiras e respectivas bacias hidrográficas como ecossistemas dependentes de águas subterrâneas. In *Formação e Ocupação de Litorais nas Margens do Atlântico—Brasil/Portugal*; Pereira, S., Freitas, J., Bergamaschi, S., Rodrigues, M., Eds.; Fundação Carlos Chagas Filho de Amparo à Pesquisa do estado do Rio de Janeiro: Rio de Janeiro, Brazil, 2014; pp. 186–213.
12. Menking, K.M.; Syed, K.H.; Anderson, R.Y.; Shafike, N.G.; Arnold, J.G. Model estimates of runoff in the closed, semiarid Estancia basin, central New Mexico, USA. *Hydrol. Sci. J.* **2003**, *48*, 953–970. [CrossRef]
13. Kim, N.W.; Chung, I.M.; Won, Y.S.; Arnold, J.G. Development and application of the integrated SWAT-MODFLOW model. *J. Hydrol.* **2008**, *356*, 1–16. [CrossRef]
14. Pulido-Velázquez, M.; Peña-Haro, S.; García-Prats, A.; Mocholi-Almudever, A.F.; Henriquez-Dole, L.; Macian-Sorribes, H.; Lopez-Nicolas, A. Integrated assessment of the impact of climate and land use changes on groundwater quantity and quality in the Mancha Oriental system (Spain). *Hydrol. Earth Syst. Sci.* **2015**, *19*, 1677–1693. [CrossRef]
15. Poeter, E.; Anderson, D. Multimodel ranking and inference in ground water modeling. *Groundwater* **2005**, *43*, 597–605. [CrossRef] [PubMed]
16. Hojberg, A.L.; Refsgaard, J.C. Model uncertainty-parameter uncertainty versus conceptual models. *Water Sci. Technol.* **2005**, *52*, 177–186. [CrossRef] [PubMed]
17. Beven, K. Towards integrated environmental models of everywhere: Uncertainty, data and modelling as a learning process. *Hydrol. Earth Syst. Sci.* **2007**, *11*, 460–467. [CrossRef]

18. Monteiro Santos, F.A.; Sultan, S.A.; Represas, P.; El Sorady, A.L. Joint inversion of gravity and geoelectric data for groundwater and structural investigation: Application to the northwestern part of Sinai, Egypt. *Geophys. J. Int.* **2006**, *165*, 705–718. [CrossRef]
19. Khalil, M.A.; Hafez, M.A.; Santos, F.M.; Ramalho, E.C.; Mesbah, H.S.; El-Qady, G.M. An approach to estimate porosity and groundwater salinity by combined application of GPR and VES: A case study in the Nubian sandstone aquifer. *Near Surf. Geophys.* **2010**, *8*, 223–233. [CrossRef]
20. Alam, K.; Ahmad, N. Determination of aquifer geometry through geophysical methods: A case study from Quetta Valley, Pakistan. *Acta Geophys.* **2014**, *62*, 142–163. [CrossRef]
21. Farzamian, M.; Monteiro Santos, F.A.; Khalil, M.A. Estimation of unsaturated hydraulic parameters in sandstone using electrical resistivity tomography under a water injection test. *J. Appl. Geophys.* **2015**, *121*, 71–83. [CrossRef]
22. Binley, A.; Hubbard, S.S.; Huisman, J.A.; Revil, A.; Robinson, D.A.; Singha, K.; Slater, L.D. The emergence of hydrogeophysics for improved understanding of subsurface processes over multiple scales. *Water Resour. Res.* **2015**, *51*, 3837–3866. [CrossRef]
23. Uhlemann, S.S.; Sorensen, J.P.R.; House, A.R.; Wilkinson, P.B.; Roberts, C.; Gooddy, D.C.; Binley, A.M.; Chambers, J.E. Integrated time-lapse geoelectrical imaging of wetland Hydrol Process. *Water Resour. Res.* **2016**, *52*, 3. [CrossRef]
24. Gonçalves, R.; Farzamian, M.; Monteiro Santos, F.A.; Represas, P.; Mota Gomes, A.; Lobo de Pina, A.F.; Almeida, E.P. Application of Time-Domain Electromagnetic Method in Investigating Saltwater Intrusion of Santiago Island (Cape Verde). *Pure Appl. Geophys.* **2017**, *174*, 4171–4182. [CrossRef]
25. Xia, J.; Miller, R.D.; Park, C.B. Estimation of near-surface shear-wave velocity by inversion of Rayleigh wave. *Geophysics* **1999**, *64*, 691–700. [CrossRef]
26. Xia, J.; Miller, R.D.; Park, C.B.; Hunter, J.A.; Harris, J.B.; Ivanov, J. Comparing shear-wave velocity profiles inverted from multichannel surface wave with borehole measurements. *Soil Dyn. Earthq. Eng.* **2002**, *22*, 181–190. [CrossRef]
27. Park, C.B.; Miller, R.D.; Xia, J. Multi-channel analysis of surface waves. *Geophysics* **1999**, *64*, 800–808. [CrossRef]
28. Park, C.B.; Miller, R.D.; Xia, J.; Ivanov, J. Multichannel analysis of surface waves (MASW)—Active and passive methods. *Lead Edge* **2007**, *26*, 60–64. [CrossRef]
29. Hayley, K.; Bentley, L.R.; Gharibi, M.; Nightingale, M. Low temperature dependence of electrical resistivity: Implications for near surface geophysical monitoring. *Geophys. Res. Lett.* **2007**, *34*, L18402. [CrossRef]
30. Hayley, K.; Bentley, L.R.; Gharibi, M. Time-lapse electrical resistivity monitoring of salt-affected soil and groundwater. *Water Resour. Res.* **2010**, *45*, W07425. [CrossRef]
31. Steelman, C.M.; Kennedy, C.S.; Capes, D.C.; Parker, B.L. Electrical resistivity dynamics beneath a fractured sedimentary bedrock riverbed in response to temperature and groundwater–surface water exchange. *Hydrol. Earth Syst. Sci.* **2017**, *21*, 3105–3123. [CrossRef]
32. Giustiniani, M.; Accaino, F.; Picotti, S.; Tinivella, U. Characterization of the shallow aquifers by high-resolution seismic data. *Geophys. Prospect.* **2008**, *56*, 655–666. [CrossRef]
33. Martorana, R.; Lombardo, L.; Messina, N.; Luzio, D. Integrated geophysical survey for 3D modelling of a coastal aquifer polluted by seawater. *Near Surf. Geophys* **2014**, *12*, 45–59. [CrossRef]
34. Foti, S.; Hollender, F.; Garofalo, F.; Albarello, D.; Asten, M.; Bard, P.-Y.; Comina, C.; Cornou, C.; Cox, B.; Giulio, G.D.; et al. Guidelines for the good practice of surface wave analysis: A product of the InterPACIFIC project. *Bull. Earthq. Eng.* **2018**, *16*, 2367–2420. [CrossRef]
35. Benjumea, B.; Gabàs, A.; Macau, A.; Bellmunt, F.; Figueras, S.; Vilà, M.; Pi, R. Combination of Geophysical Techniques to Characterize Sediments (Ebro Delta, Spain). In Proceedings of the Near Surface Geoscience 2016—22nd European Meeting of Environmental and Engineering Geophysics, Barcelona, Spain, 4–6 September 2016; European Association of Geoscientists & Engineers: Houten, The Netherlands, 2016. [CrossRef]
36. Billy, J.; Baudouin, V.; Portal, A.; Deparis, J.; Bitri, A.; Garcin, M. An Innovative Approach for a Comprehensive Characterization of Coastal Dune Systems through Internal Architecture and the Associated Intrinsic Geophysical Properties. *J. Coast. Res.* **2020**, *95*, 387–391. [CrossRef]
37. Glazer, M.; Dobiński, W.; Marciniak, A.; Majdański, M.; Błaszczyk, M. Spatial distribution and controls of permafrost development in non-glacial Arctic catchment over the Holocene, Fuglebekken, SW Spitsbergen. *Geomorphology (Amst)* **2020**, *358*, 107128. [CrossRef]

38. Telford, W.M.; Geldart, L.P.; Sheriff, R.E. *Applied Geophysics*, 2nd ed.; Cambridge University Press: New York, NY, USA, 1990.
39. Alcalá, F.J.; Custodio, E. Atmospheric chloride deposition in continental Spain. *Hydrol. Process.* **2008**, *22*, 3636–3650. [CrossRef]
40. Ramos, R. Contribuição dos Métodos Geofísicos Para o Modelo Evolutivo da Região de Santo André Desde o último Máximo Glaciário. Ph.D. Thesis, University of Lisbon, Lisbon, Portugal, 2013. Available online: http://hdl.handle.net/10451/10795 (accessed on 1 May 2020).
41. Trigo, R.M.; Pozo-Vázquez, D.; Osborn, T.J.; Castro-Díez, Y.; Gámiz-Fortis, S.; Esteban-Parra, M.J. North Atlantic Oscillation influence on precipitation, river flowand water resources in the Iberian Peninsula. *Int. J. Climatol.* **2004**, *24*, 925–944. [CrossRef]
42. APA. *Plano de Gestão das Bacias Hidrográficas Integradas nas Regiões Hidrográficas 6 e 7*; Agência Portuguesa do Ambiente: Lisboa, Portugal, 2012; Volume 1.
43. LNEG. *Geological Map of Portugal, Scale 1:50,000; Sheet nº 42-A, Grândola; National Laboratory of Energy and Geology of Portugal, Memory and Maps*. 2011. Available online: http://geoportal.lneg.pt/geoportal/egeo/DownloadCartas/ (accessed on 28 February 2020).
44. Dias, R.; Araújo, A.; Terrinha, P.; Kullberg, J.C. (Eds.) *Geologia de Portugal*; Escolar Editora: Lisboa, Portugal, 2013; Volume 2, p. 1624.
45. Quesada, C.; Oliveira, J.T. (Eds.) *The Geology of Iberia: A Geodynamic Approach*; Springer Nature Switzerland AG: Cham, Switzerland, 2020. [CrossRef]
46. Almeida, C.; Mendonça, J.J.; Jesus, M.R.; Gomes, J. *Sistemas Aquíferos de Portugal Continental*; Centro de Geologia da FCUL, Instituto da Água: Lisboa, Portugal, 2000; p. 649.
47. Freitas, M.C.; Andrade, C.; Rocha, F.; Tassinari, C.; Munhá, J.M.; Cruces, A.; Vidinha, J.; Silva, C.M. Late glacial and Holocene environmental changes in Portuguese coastal lagoons 1: The sedimentological and geochemical records of the Santo André coastal area. *Holocene* **2003**, *13*, 433–446. [CrossRef]
48. Cearreta, A.; Cachão, M.; Cabral, M.C.; Bao, R.; Ramalho, M.J. Lateglacial and Holocene environmental changes in Portuguese coastal lagoons 2: Microfossil multiproxy reconstruction of the Santo André coastal area. *Holocene* **2003**, *13*, 447–458. [CrossRef]
49. Pires, A.R.M. Morfodinâmica da Barra de Maré Efémera da Lagoa de Santo André. Master's Thesis, University of Lisbon, Lisbon, Portugal, 2011. Available online: http://hdl.handle.net/10451/8451 (accessed on 1 May 2020).
50. Lobo-Ferreira, J.P.C.; Oliveira, M.M.; Moinante, M.J.; Leitão, T.E.; Novo, M.E.; Moreira, P.E.; Henriques, M.J. *Caracterização dos Recursos Hídricos Subterrâneos da Área Abrangida pelo Plano de Bacia Hidrográfica do Rio Sado.Anexo 4 Recursos Hídricos Subterrâneos. Relatório Provisório da Fase I*; Laboratório Nacional de Engenharia Civil: Lisboa, Portugal, October 1999; p. 135.
51. Monteiro, J.P.; Chambel, A.; Martins, J. *Conceptual and Numerical Flow Model of the Sines Aquifer System (Alentejo, South Portugal). International Groundwater Symposium*; International Association of Hydraulic Engineering and Research (IAHR): Istanbul, Turkey, 2008; p. 38 (abstract) and p. 9 (CD-Rom).
52. Alcalá, F.J.; Cantón, Y.; Contreras, S.; Were, A.; Serrano-Ortiz, P.; Puigdefábregas, J.; Solé-Benet, A.; Custodio, E.; Domingo, F. Diffuse and concentrated recharge evaluation using physical and tracer techniques: Results from a semiarid carbonate massif aquifer in southeastern Spain. *Environ. Earth Sci.* **2011**, *63*, 541–557. [CrossRef]
53. Kim, J.; Supper, R.; Tsourlos, P.; Yi, M. Four-dimensional inversion of resistivity monitoring data through L*p* norm minimizations. *Geophys. J. Int.* **2013**, *195*, 1640–1656. [CrossRef]
54. Loke, M.H. *Rapid 2-D Resistivity & IP Inversion Using the Least-squares Method*; Geotomo Software: Penang, Malaysia, 2012.
55. Chapman, T.G. A comparison of algorithms for stream flow recession and baseflow separation. *Hydrol. Process* **1999**, *13*, 701–714. [CrossRef]
56. IPCC. *Climate Change 2007: Impacts, Adaptation and Vulnerability. Contribution of Working Group II to the Fourth Assessment Report of the Intergovernmental Panel on Climate Change*; Cambridge University Press: Cambridge, UK, 2007.
57. Navarro, M.; Vidal, F.; Enomoto, T.; Alcalá, F.; García-Jerez, A.; Sánchez, F.J.; Abeki, N. Analysis of the weightiness of site effects on reinforced concrete (RC) building seismic behavior. The Adra town example (SE Spain). *Earthq. Eng. Struct. D* **2007**, *36*, 1363–1383. [CrossRef]

58. García-Jerez, A.; Navarro, M.; Alcalá, F.J.; Luzón, F.; Pérez-Ruiz, J.A.; Enomoto, T.; Vidal, F.; Ocaña, E. Shallow velocity structure using joint inversion of array and h/v spectral ratio of ambient noise: The case of Mula town (SE of Spain). *Soil Dyn. Earthq. Eng.* **2007**, *27*, 907–919. [CrossRef]
59. Martínez-Pagán, P.; Navarro, M.; Pérez-Cuevas, J.; Alcalá, F.J.; García-Jerez, A.; Vidal, F. Shear-wave velocity structure from MASW and SPAC methods. The case of Adra town, SE Spain. *Near Surf. Geophys.* **2018**, *16*, 356–371. [CrossRef]
60. Alcalá, F.J.; Espinosa, J.; Navarro, M.; Sánchez, F.J. Propuesta de división geológica de la localidad de Adra (provincia de Almería). Aplicación a la zonación sísmica. *Rev. Soc. Geológica España* **2002**, *15*, 55–66.
61. Mitchell, J.K.; Soga, K. *Fundamentals of Soil Behaviour*; Wiley: London, UK, 2005.
62. Zimmer, M.A.; Prasad, M.; Mavko, G.; Nur, A. Seismic velocities of unconsolidated sands: Part 1—Pressure trends from 0.1 to 20 MPa. *Geophysics* **2007**, *72*, E1–E13. [CrossRef]
63. Martínez-Pagán, P.; Navarro, M.; Pérez-Cuevas, J.; Alcalá, F.J.; García-Jerez, A.; Sandoval-Castaño, S. Shear-wave velocity based seismic microzonation of Lorca city (SE Spain) from MASW analysis. *Near Surf. Geophys.* **2014**, *12*, 739–749. [CrossRef]
64. McGann, C.R.; Bradley, B.A.; Cubrinovski, M. Investigation of shear wave velocity depth variability, site classification, and liquefaction vulnerability identification using a near-surface Vs model of Christchurch, New Zealand. *Soil Dyn. Earthq. Eng.* **2017**, *92*, 692–705. [CrossRef]
65. Alcalá, F.J.; Custodio, E. Natural uncertainty of spatial average aquifer recharge through atmospheric chloride mass balance in continental Spain. *J. Hydrol.* **2015**, *524*, 642–661. [CrossRef]
66. Willmott, C.J.; Matsuura, K. Advantages of the mean absolute error (MAE) over the root mean square error (RMSE) in assessing average model performance. *Clim. Res.* **2005**, *30*, 79–82. [CrossRef]
67. Chai, T.; Draxler, R.R. Root mean square error (RMSE) or mean absolute error (MAE)?—Arguments against avoiding RMSE in the literature. *Geosci. Model. Dev.* **2014**, *7*, 1247–1250. [CrossRef]
68. Hyndman, R.J.; Koehler, A.B. Another look at measures of forecast accuracy. *Int. J. Forecast.* **2006**, *22*, 679–688. [CrossRef]
69. Moriasi, D.N.; Wilson, B.N.; Douglas-Mankin, K.R.; Arnold, J.G.; Gowda, P.H. Hydrologic and water quality models: Use, calibration, and validation. *Trans. ASABE* **2012**, *55*, 1241–1247. [CrossRef]
70. Lin, L.; Hedayat, A.S.; Wu, W. *Statistical Tools for Measuring Agreement*; Springer: New York, NY, USA, 2012.
71. Frid, V.; Averbach, A.; Frid, M.; Dudkinski, D.; Liskevich, G. Statistical Analysis of Resistivity Anomalies Caused by Underground Caves. *Pure Appl. Geophys.* **2017**, *174*, 997–1012. [CrossRef]
72. Frid, V.; Sharabi, I.; Frid, M.; Averbakh, A. Leachate detection via statistical analysis of electrical resistivity and induced polarization data at a waste disposal site (Northern Israel). *Environ. Earth Sci.* **2017**, *76*, 233. [CrossRef]
73. Vásconez-Maza, M.D.; Martínez-Segura, M.A.; Bueso, M.C.; Faz, A.; García-Nieto, M.C.; Gabarrón, M.; Acosta, J.A. Predicting spatial distribution of heavy metals in an abandoned phosphogypsum pond combining geochemistry, electrical resistivity tomography and statistical methods. *J. Hazard. Mater.* **2019**, *374*, 392–400. [CrossRef] [PubMed]

applied sciences

MDPI

Article

Geophysical Characterization of Aquifers in Southeast Spain Using ERT, TDEM, and Vertical Seismic Reflection

Javier Rey [1,*], Julián Martínez [2], Rosendo Mendoza [2], Senén Sandoval [3], Vladimir Tarasov [4], Alex Kaminsky [5], M. Carmen Hidalgo [1] and Kevin Morales [1]

1 Departamento de Geología, EPS de Linares y CEACTEMA, Universidad de Jaén, Campus Científico Tecnológico, 23700 Linares, Jaén, Spain; chidalgo@ujaen.es (M.C.H.); kamr0001@red.ujaen.es (K.M.)
2 Departamento de Ingeniería Mecánica y Minera, EPS de Linares y CEACTEMA, Universidad de Jaén, Campus Científico Tecnológico, 23700 Linares, Jaén, Spain; jmartine@ujaen.es (J.M.); rmendoza@ujaen.es (R.M.)
3 Everest Geophysics SL, 28270 Colmenarejo, Madrid, Spain; senen@everestgeophysics.com
4 Geophysicist in ElGeo Ltd., Petrovskaya Kosa, 1, 190000 St. Petersburg, Russia; vladtar@yandex.ru
5 Zond Software Ltd., 8010 Paphos, Cyprus; zondgeo@gmail.com
* Correspondence: jrey@ujaen.es

Received: 29 July 2020; Accepted: 17 October 2020; Published: 21 October 2020

Abstract: We assess the effectiveness of complementary geophysical techniques to characterize a Jurassic dolomite confined aquifer at Loma de Ubeda, Spain. This aquifer, which is penetrated by wells in the 100–600-m depth range, is confined by Triassic clays (bottom) and Miocene marls (top). The Jurassic dolomite is characterized by prominent seismic reflectors of high amplitude. Thus, it is readily differentiated from the low-amplitude reflectors of the confining clay-rich Triassic and Miocene materials. Electrical resistivity tomography (ERT) allowed us to detail the characteristics of the aquifer up to a maximum depth of 220 m. Lateral changes in facies and small faults have been identified using ERT. Time-domain electromagnetic (TDEM) is an excellent complement to the two above-mentioned techniques in order to widen the analyzed depth range. We acquire TDEM data with different configurations at multiple study sites while simultaneously varying measurement parameters. In doing so and by comparing the effectiveness of these different configurations, we expand the use of TDEM for aquifer characterization.

Keywords: Jurassic dolomite aquifer; seismic reflection; electrical resistivity tomography; time-domain electromagnetic; loma de Úbeda; Spain

1. Introduction

Geophysical prospecting techniques for groundwater exploration have evolved considerably in recent years. The automatic data acquisition systems, the availability of powerful computers, and the development of two- and three-dimensional modeling software have greatly improved the resolution of complex geological models [1–4].

A great diversity of geophysical methods (gravimetric, magnetic, electrical, electromagnetic, and seismic) can contribute to the groundwater exploration. Electrical, electromagnetic, and seismic methods are the most efficient techniques in this kind of setting and they are more cost-effective.

Electrical methods are particularly suitable for groundwater research because hydrogeological parameters, such as porosity and permeability, can be correlated to electrical resistivity values. In addition, they are cheap and easy to implement techniques [5]. Electromagnetic sounding techniques, on the other hand, have not been used as extensively as electrical methods for groundwater research because the equipment is more expensive and the interpretation methods are more complicated.

The use has focused on detecting the interface between fresh groundwater and saline seawater intrusion into coastal aquifers [6]. These two techniques have been used recently in an unconfined sandstone aquifer in semiarid southwestern Niger [7]. This study is a good example of the efficiency of using complementary geophysical techniques.

Herein, we analyze the effectiveness of electrical (electrical resistivity tomography, ERT), electromagnetic (time-domain electromagnetic, TDEM), and reflection seismic techniques to characterize dolomite aquifers. Our study area (Figure 1) not only has detailed surface geological information, but also numerous pumping boreholes and seismic lines, which were employed to verify and correlate the new geophysical data collected in this research.

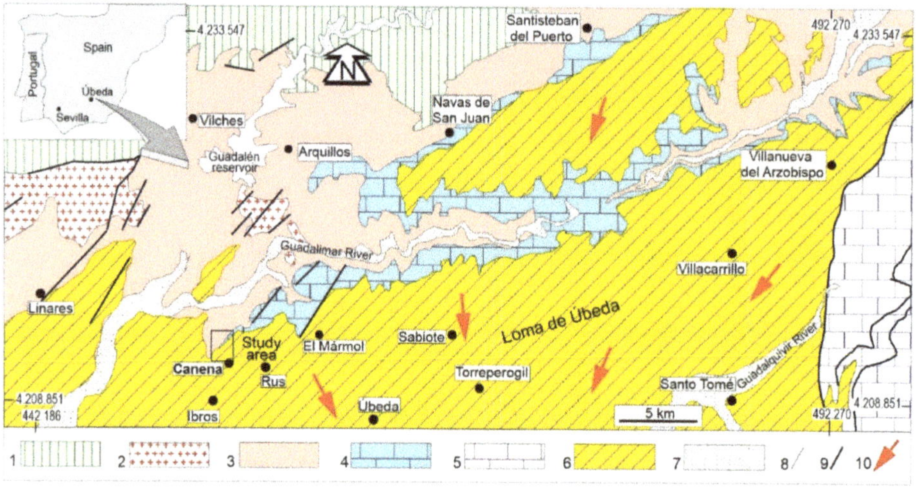

Figure 1. Regional hydrogeological map. The location of the study area is shown. Legend: (1) Paleozoic basement (Phyllites). (2) Paleozoic basement (Granite). (3) Triassic clays from Chiclana de Segura Formation (Impervious base). (4) Jurassic tabular dolomites (Aquifer). (5) Prebetic Jurassic dolomites (Aquifer). (6) Late Miocene marls with detrital levels. (7) Pliocene–Quaternary clays, marls, and conglomerates. (8) Undifferentiated stratigraphic contact. (9) Undifferentiated tectonic contact. (10) Groundwater flow paths.

The economy of Loma de Úbeda area in Southern Spain is based almost exclusively on the cultivation of olive groves. Groundwater exploitation has increased olive production and ensured the survival of crops during the severe droughts that affect this region.

Until the late 1980s, only a Miocene shallow aquifer was exploited, with pumping flows on the order of 1 L/s. In the early 1990s, the Loma de Úbeda Jurassic dolomite aquifer began to be exploited at depths between 100 m and 600 m, with pumping rates exceeding 40 L/s in some cases [8]. Numerous deep boreholes were drilled, many of them without the mandatory permits from the Water Authorities. Estimated irrigation in 2005 exceeded 35 hm^3/year over an irrigation area of 24,050 ha of olive groves [9]. The aquifer has an average annual deficit close to 13.3 hm^3 and it is clearly overexploited [10]. We applied different geophysical techniques to make comparisons with existing boreholes and surface geologic data. This allowed us to verify the effectiveness of each technique to investigate this aquifer. We emphasized TDEM and we used different configurations to find the optimal parameters. The different techniques used here can be applied to investigate other carbonate aquifers showing similar behavioral.

2. Description of the Study Region

2.1. Geological Setting

Two geological units can be differentiated in the study area: the Paleozoic basement and the post-Hercynian sedimentary cover. The first unit is dominated by intensely folded phyllites intruded by a granitic batholith (Figure 1: 1,2). Subhorizontal to gently dipping, the post-Hercynian sedimentary cover unconformably overlies the basement. It is composed of Triassic, Jurassic, and Neogene formations (Figure 1). Sparse Quaternary alluvial sediments are also present.

The Triassic strata belongs to the Chiclana de Segura Formation [11,12], which extensively outcrops at the north of the Guadalimar River (Figure 1: 3). This formation varies in thickness (50–400 m) and is essentially horizontal or dips gently southward. In these rocks, their reddish hues and the presence of shales and sandstones are characteristic (Figure 2). The lower part of the Triassic series is mostly composed by sandstones levels. Towards the upper-middle part, the number and thick of clayey facies increases, and gypsum and salt levels intercalations appear (Figure 2).

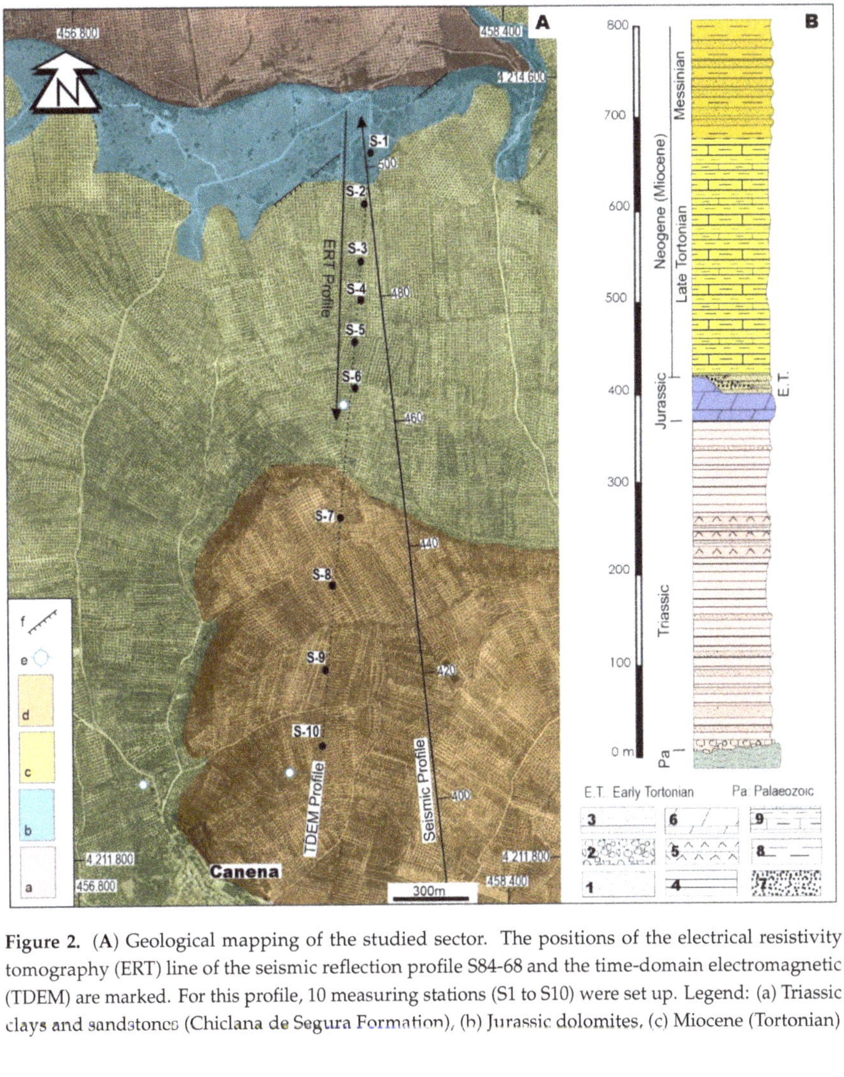

Figure 2. (**A**) Geological mapping of the studied sector. The positions of the electrical resistivity tomography (ERT) line of the seismic reflection profile S84-68 and the time-domain electromagnetic (TDEM) are marked. For this profile, 10 measuring stations (S1 to S10) were set up. Legend: (a) Triassic clays and sandstones (Chiclana de Segura Formation), (b) Jurassic dolomites, (c) Miocene (Tortonian)

marl and marly limestones, (d) Miocene (Messinian) sandstones and marls, (e) boreholes, and (f) normal faults. (**B**) Synthetic stratigraphic column. Legend as (1) phyllites, (2) rudites, (3) sandstones, (4) clays, (5) gypsum, (6) dolomites, (7) gravel-rich sediments, (8) marls, and (9) marls and marly limestones.

In the Guadalimar River valley, the Jurassic dolomites directly outcrop over the Triassic materials in apparent stratigraphic continuity (Figure 1: 4). They appear subhorizontal or with small dips, generally towards the south–southeast. These are strongly brecciated and dolomitized carbonates, thus causing an important secondary porosity that must affect the aquifer storability. The thickness, highly variable in the 0–70-m range, generally decreases towards the west. Although there are no biostratigraphic criteria, this lithological unit has been attributed to the Lower Jurassic following facies and stratigraphic position criteria [12,13].

Jurassic strata dip slightly towards the south–southeast and they folded gently on NE–SW axes [5]. On the other hand, two NE–SW and NNW–SSE fault directions in both the basement and the dolomites. The orientations of theses faults overall influenced large-scale regional patterns of early Tortonian sedimentation [14–16]. Jurassic strata pinch out towards the west–northwest. Several pumping boreholes near Rus-Canena document this trend (Figure 1). West of Ibros, Jurassic strata are absent and the Miocene strata rest directly over the Triassic ones. The disappearance of Jurassic dolomites also marks the western limit of the aquifer [8].

Three Miocene units are differentiated (Figure 1: 6): at the base, the Early Tortonian, which crops out locally, filling a small trough and pinching out laterally (Figure 2). These are gravel-rich sediments with dolomite fragments and siliceous sands, with little or no cementation, alternating with marls [14]. Late Tortonian strata either directly overlie the Early Tortonian ones or they overlie the Jurassic dolomites. They average 200 m in thickness, attaining a maximum thickness of 500 m. These strata consist chiefly of marls and marly limestone, but there are also minor calcarenite intercalations [13,17]. Messinian strata with more abundant calcarenites overlie the Late Tortonian strata [13,17] (Figure 2).

2.2. *Hydrogeological Context*

The Jurassic dolomites constitute an important hydrogeological unit underlying nearly 800 km^2 (Figure 1: 4). This aquifer behaves as unconfined in the northwestern portion of the study area (Figure 2). Towards the south, it is confined by Miocene marls, while it is limited by dipping faults of the Prebetic units to the east, by the olistostromic units of the Guadalquivir River depression to the south and west it pinches out [8,18]. In the western sector, the general groundwater flow path is NNW to SSE. Fracture systems compartmentalize the aquifer, changing piezometry levels and groundwater flow paths from place to place [18].

Diffuse recharge occurs by direct infiltration from precipitation whereas preferential recharge from the Guadalimar River streamflow infiltration occurs in those northern sectors where riverbed intersects the carbonate formations (Figure 1). It is also worth considering the existence of lateral transferences from eastern Prebetic units (Figure 1: 5).

The exploitation of the confined sector of the aquifer began in the early 1990s. The mechanical drilling methods before that data did not exceed 100 m depth from shallow Miocene strata only. The Jurassic dolomite aquifer has been exploited with pumping rates between 15 and 40 L/s only since 10 years ago thanks to deeper boreholes with depths in the 200–600-m range. Today, the pumping rate is much lower given the overexploitation that the aquifer unit suffers.

Different groundwater chemical facies can be differentiated. Shallow boreholes (dolomites pumped at 200–300 m) generally yield slightly mineralized waters. Salinity increases considerably with depth, and deeper chemical facies pass from magnesium-calcium sulfate, sodium bicarbonate-sulfate, sodium bicarbonate, to sodium chloride [18]. Chemical reduction processes in the deeper confined parts [18] and mixing of groundwater from Triassic, Jurassic, and Miocene aquifers have been described [9,19].

3. Methods

3.1. Seismic Reflection

Chevron Oil Company of Spain acquired seismic profiles in southeastern Spain in 1983 and 1984. These surveys used arrays of 18 geophones (two chains of nine geophones per trace) at intervals of 20 m. The processing flow followed a standard processing sequence, including migration. The information was processed by the 'Compagnie Générale de Géophysique'. The integrated information of the different seismic lines, together with the direct data from the pumping boreholes, allowed a first reconstruction of the 3D geometry of the Jurassic dolomite unit [8].

This study focuses on seismic profile S84-68 (Figure 2). A correction speed of 2000 m/s was applied to the two-way travel time/depth transformation in Miocene strata. This is the wave velocity that is marked in the processing of the line. The thicknesses that result applying this wave velocity are correlated in pumping boreholes.

3.2. Electrical Resistivity Tomography

This geophysical prospecting technique consists of determining the distribution of the electrical resistivity of the subsoil from a very large number of measurements collected from the ground surface. The electrical resistivity quantifies how a material resists or conducts electric current [20,21]. The different electrical behavior of geological materials allows us to obtain 2D resistivity models, making ERT one of the most effective non-destructive tools to study and characterize subsurface discontinuities [20,22]. In recent years, this technique has attained widespread success in stratigraphic, hydrogeological, and environmental studies [3,23–29].

ERT involves the installation of numerous electrodes along a line (profile), with a given separation that determines the resolution and prospecting depth to be reached. A smaller separation between electrodes increases resolution, whereas a larger separation increases prospecting depth [22]. To make one measurement, only four electrodes are needed. Two of them act as 'current electrodes' and the other two act as 'potential electrodes'. The way in which the electrodes are selected is termed 'electrode configuration'. In stratigraphic studies, the Wenner–Schlumberger configuration is usually selected [30,31]. This configuration has good behavior and stability against resistivity changes, both vertical and horizontal ones, so it is useful for the investigation of horizontal or slightly inclined layers that can present lateral changes of facies and/or verticalized structures, as in our case [30,31].

Electrodes are connected to the measuring equipment (resistivity-meter), and through a predefined sequence, the groups of electrodes are selected. For each electrode quadrupole, a voltage and an intensity measurement are made. With these two readings, the 'apparent' resistivity of the ground is calculated, which is attributed to a certain geometric point in the subsoil.

We used a multi-channel, multi-electrode DC resistivity meter system manufactured by Deutsche Montan Technologie (RESECS model). The apparent resistivity values measured in the field were inverted to obtain electrical resistivity models by using the RES2DINV software [2]. This software applies a least-squares method with a damped smoothing, modified with the quasi-Newton optimization method. This inversion method constructs a subsoil model using rectangular prisms and determines the resistivity values for each of them, minimizing the difference between the observed and calculated apparent resistivity values [2,32]. We produced a NW–SE electrical tomography profile, with a total length of 1110 m using 112 electrodes spaced at 10 m (Figure 2).

3.3. Time-Domain Electromagnetic

The operating principle of the time-domain electromagnetic (TDEM) method is to circulate an electric current through a transmitter coil (usually square in shape) for short time intervals. When the current flow is abruptly interrupted, a magnetic field is produced that induces, according to Faraday's law, a variable electric current in the subsurface, which in turn generates a transient secondary magnetic field. These currents flow in closed paths and migrate at depth, decreasing in intensity over time.

Changes in the secondary magnetic field over time induce a transient voltage in the receiving coil. The shape of the decay of this voltage provides information on the conductivity distribution of the subsoil, which can be used to characterize it [21,33–35]. TDEM has been typically applied to finding mineral deposits [36–38], to investigate groundwater bodies [35,39], and to analyze sedimentary basins [40,41]. In recent years, this method has also been used in environmental studies [42,43] and to characterize marine intrusion [44,45].

The maximum prospecting depth depends on the decay time of the signal from the current cut, the current intensity, the loop size, the signal frequency, and the conductivity of the subsoil [33–35]. Although there are various measurement configurations, the most common one is to place the (smaller) receiving coil in the center of the (much larger) transmitting coil (central-loop configuration). Another option is to use the same coil to carry out both functions (single-loop configuration).

We used a TEM measuring system developed and produced by ELGEO Research & Production Company (AIE-2 model). It is a device with a maximum output power of 200 W and a current intensity of up to 10 A. The TDEM receiver is based on a 16-bit analogue-to-digital converter and a signal processor that provides an analogue-to-digital conversion immune to input voltage noise and real-time signal pre-processing. It offers a range of measurements between 5 μs and 10 μs, with an input voltage of 5–20 V with signal compensation.

In the field campaign, a total of 10 TDEM measurement stations were acquired along a north–south profile. The points located in Figure 2 refer to the center of the measuring stations for square loops of 200 m on each side. At each of these points, measurements were collected with central-loop and single-loop configurations, and the measurement parameters (intensity, time, voltage) were varied, which allowed us to compare the effectiveness of the different setups. To reach prospecting depths on the order of 400 m, transmission loops of 200 m × 200 m are needed. A 6-mm^2 section copper wire was used. This (larger than usual) wire decreased the resistance and thereby allowed it to obtain a higher effective current intensity. When the central-loop configuration was used, the receiving coil had a dimension of 20 m × 20 m, using a 5-turn cable and a ×10 amplifier. For the visualization and editing of the different curves, the TEMBIN software was used. The modeling and inversion processes were performed with ZondTEM1D and ZondTEM2D software http://zond-geo.com/english/zond-software/electromagnetic-sounding/zondtem1d/.

4. Results and Discussion

4.1. Seismic Reflection

The sedimentary cover sequence at the top of the seismic profiles produces reflectors of moderate amplitude, which we interpreted as Miocene sandstones and marls. The seismic response of the Jurassic dolomites consists of prominent reflectors of high amplitude and continuity. This different response allows us to trace both the upper and lower boundaries of the Jurassic aquifer. Under the Jurassic unit, locally discontinuous reflectors of low amplitude are associated to the Late Triassic unit, which contains clays and evaporates (gypsum and halite). At the base of the Triassic unit, continuous levels of greater amplitude are detected, which we have correlated with the basal, thick Triassic sandstones levels. The acoustic material is the Paleozoic basement, which is characterized by chaotic features and disorganized seismic facies.

Different criteria to calculate the stratigraphic units' thickness were used. A criterion, namely the measured thicknesses of stratigraphic units, was measured in the field. The other criterion was velocity values included with the profiles. These two criteria generally coincided. Finally, surveys carried out in the area provide some direct assessment of local hydrogeological conditions.

The seismic profile selected in this study (Figures 2 and 3A) shows Jurassic dolomites dip approximately 20° S–SE. The unit outcrops in the northern sector and occupies deeper positions in the southernmost sectors. The resolution of the profile is very low near the land surface, so the dolomites are not readily identifiable to the north (Figure 3A).

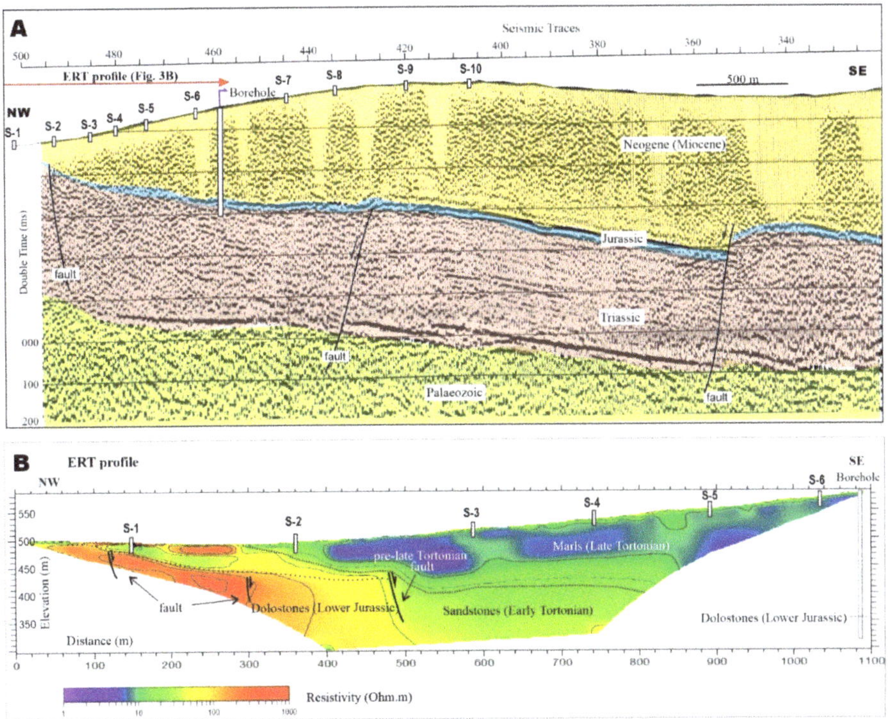

Figure 3. Seismic reflection profile and interpretation (see location in Figure 2). The positions of the pumping boreholes and the TDEM (S1 to S10) measuring stations are indicated. M.a.s.l. (metres above sea level) (**A**). Electrical resistivity tomography (ERT) profile (**B**).

Using a correction speed of 2000 m/s, the dolomites can be identified at 200 m depth near the pumping boreholes, as pointed out in Figure 3A. This depth is confirmed from borehole drilling information. Under the dolomites, higher-amplitude reflectors in the lower part of the Triassic unit correspond to sandstones. Some of the faults detected with ERT techniques affecting the sedimentary cover are also identified by seismic methods.

4.2. Electrical Resistivity Tomography

The ERT profile was acquired at the northernmost end (Figure 2), just where the dolomite aquifer is too shallow to be discerned in the seismic profile (Figure 3A). Figure 3B shows the ERT profile, which reaches a depth of 220 m in the maximum penetration zone. From geoelectrical records, three resistivity ranges are differentiated. The first is characterized by high resistivity values in the 100–500 Ωm range and corresponds to the Jurassic dolomite aquifer. These geoelectrical facies outcrops at the northern profile end and dip gently towards the south–southeast. The ERT results trace this facies up to 500 m where it suddenly ends, and this could be explained by the presence of a normal fault. In the local outcrops, normal faults affecting the Jurassic dolomites have been detected. In the same ERT profile, normal faults affecting this unit can be deduced. Field observations and geological maps (Figure 2) corroborate these interpretations. This Jurassic aquifer crops out in the northern sector and it is also found under the Miocene strata in the southern sector.

Over the dolomite aquifer, the ERT profile also shows a discontinuous unit of variable thickness characterized by resistivity values in the 40–80 Ωm range. This electrical behavior of this second unit may correspond to Early Tortonian sandy and gravel-rich sediments with dolostone fragments.

The presence of pre-Late Tortonian faults determines the accommodation space for these syntectonic facies (Figure 2), as interpreted in previous studies [14]. In the northernmost sector, some of these facies having similar resistivity values could also correspond to weathered dolomites, conferring some uncertainty in the interpretation.

A third unit characterized by resistivity values in the 5–20 Ωm range can be identified at shallow depth from 400 m onwards. Based on its electrical behavior, this unit is interpreted to be Tortonian marls and marly limestones, which have great thickness in the southern sector. As can be deduced from the geological map (Figure 2), the Messinian facies are absent in the ERT profile.

The resistivity of the different lithologies is related to different physical parameters of the geological materials, such as texture, pore fluids, density, etc. [21]. As deduced, if knowledge of the subsoil is scarce, results are more uncertainty. In contrast, if the information available is abundant, these techniques can have multiple applications, for instance to identify lateral changes of facies, thickness variations of stratigraphic units, water-table positions, or degree of saline contamination of an aquifer.

4.3. Time-Domain Electromagnetic

For each of the measurement stations of the TDEM profile, the configuration parameters of loops, current injection, voltage, or measurement time were varied. Figure 4A shows the different induced voltage decay curves measured at station 1 (S1) for a single-loop configuration and obtained using the TEMBIN software.

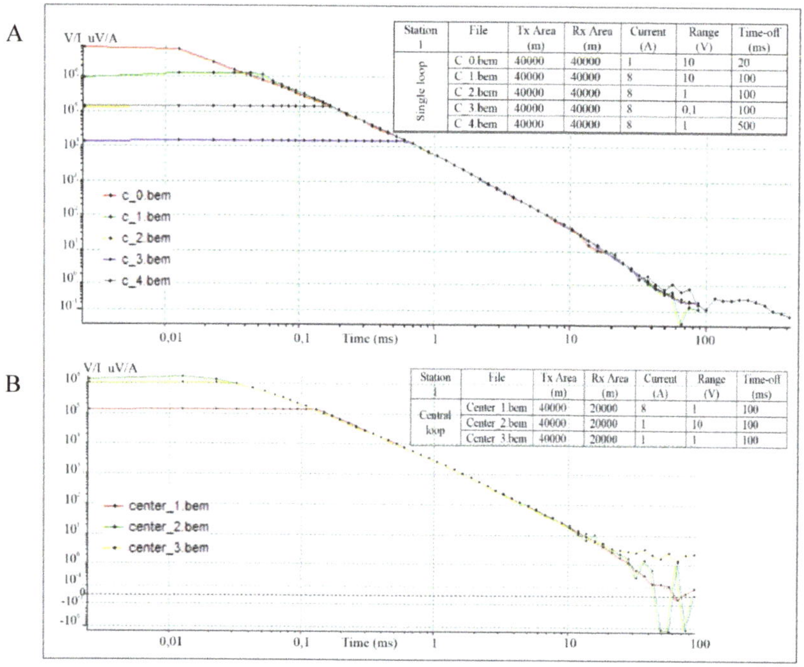

Figure 4. Different induced voltage curves as a function of time, all collected at station S1. Different measurement parameters were used in each of the curves (variations in Current, Voltage, Time). (**A**) The curves represent a single-loop configuration so that the transmitter area (Tx) and the receiver area (Rx) coincide (200 m × 200 m = 40,000 m^2). (**B**) The curves represent a central-loop configuration, where the same transmitter loop is maintained, but the receiver loop (Rx) has dimensions of 20 m × 20 m; in this case, a five-wire cable and an amplifier of ×10 were used, so the effective area of Rx in this case is 20 × 20 × 5 × 10 = 20,000 m^2.

In each of the tests, the ramp time is directly proportional to the current used [46]. In our study, according to the technical specifications of the equipment, the ramp time was 12 µs or 30 µs, depending on whether the injected current was 1 or 8 A (for loops of 200 m × 200 m). In station S1, when comparing the curves 'C_0.bem' and 'C_1.bem' (Figure 4A), currents of low amperage led the plateau at the top of the curve to be lower, and this increases with increasing amperage. Therefore, the information of the surface part decreases with the amplitude of the current. However, the capacity for in-depth investigation increases with the magnetic moment ($I * A$). According to Spies [34]:

$$d \sim 0.55\left(\frac{I * A}{\sigma * Nr}\right)^{\frac{1}{5}}$$

where: d is depth in m, I is current of the transmitter in Amps, A the effective area of the transmitter in m^2, σ is electrical conductivity of the medium in S/m, and Nr is noise level in V/m^2.

By keeping the surface of the transmitter loop constant (Tx = 200 m × 200 m), the magnetic moment and the penetration capability remain proportional to the current, as deduced when curves 'C_0.bem' and 'C_1.bem' are compared (Figure 4A). Therefore, high currents are used even if information is lost near the top of the profile. Such information from very shallow depths is not relevant to the hydrogeological interpretation.

Another important aspect to consider is the voltage range. We compare the response with the measurements at 0.1, 1, and 10 V at station S1 (Figure 4A). The superposition of the different curves allows us to deduce that greater stability is achieved in the depth of the induced voltage curve using low voltages as 0.1 V in the 'C_3.bem' curve of Figure 4A.

Another factor analyzed was the measurement time window; longer time windows produce larger depths of investigation [34]. In our case, different time-offs at station S1 (20, 100, and 500 ms) were used. When the 'C_2.bem' and 'C_4.bem' curves are compared, times greater than 100 ms are observed, the noise level increases considerably, and the voltage values are extremely low, so time intervals in this range should be ignored due to their low reliability.

All the aforementioned curves were obtained with a single loop. The readings were repeated at station S1 (Figure 4B) for the measurements with the different set-ups but using a central-loop device. As in the previous case, the data for research at deeper levels show that the magnetic moment must be increased (if the area remains constant, it will be proportional to the increase in current) and the voltage must be decreased. The measurement time was held constant at 100 ms.

In all the stations indicated in the profile (Figure 2), the previously mentioned parameters were used. Figure 5 shows the fit between the induced voltage curves measured in the field (dots) and those generated (curves) by the resistivity model (thick red line). The figure also shows the apparent resistivity obtained as a function of time in the first and last of these stations (S-1 and S-10). For each of these curves, the fit between the field and the model (by the ZONDTEM1D software) lines is observed.

The curves observed in the field at station S1 and the modeled ones (Figure 2) are represented and compared when we use a central-loop device ('Center 1' in Figure 5A) and a single-loop device ('C_3' in Figure 5B). First, the good fit of the curves stands out in both cases, with a root mean square (RMS) error of 2.7% and 3.3%, respectively. After the data inversion in both cases, similar results in the resistivity/depth curves are observed. In both cases, at shallow depths, there is an increase in resistivity values that correlates with the Jurassic dolomite aquifer, unit that outcrop in nearby sectors (Figure 2). Under this unit, a set of more conductive facies appears that would correspond to the Triassic clays. Within the Triassic unit, generally in the upper part, the increase in resistivity can be associated with the increase in evaporitic layers (Figure 5A,B). Below 400 m, the sharp increase in resistivity is associated with phyllites of the Paleozoic basement. This interpretation is corroborated by the information of the thicknesses observed in the nearby outcrops (Figure 2B) and to the information provided by the seismic profile in Figure 3A. Therefore, TDEM curves could determine the thickness of a geological unit accurately. However, TDEM curves do not allow to establish the correct resistivity value for each level, as other authors have already pointed out. In this study, Figure 5 represents resistivity values after inversion; their abrupt increases or decreases allow deducing the presence of different lithologies. Nevertheless, the modeled resistivity values do not correspond to the real resistivity of the proposed lithologies.

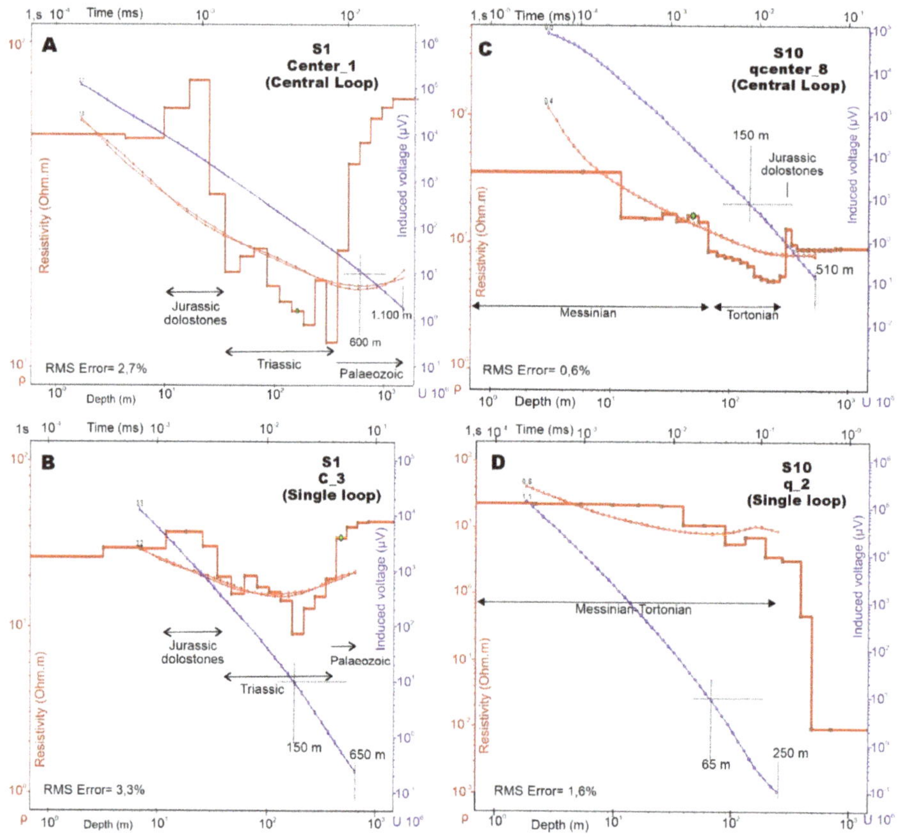

Figure 5. Induced voltage curves (blue line) and apparent resistivity (red line) as a function of time. For each of them, the fit between the field and the model (by the ZONDTEM1D software) lines is indicated. The interpretation after the inversion of the data is also included; red straight lines represent the resistivity as a function of the depth. In charts A and B, data collected at station S1 (see Figure 2) using two different devices are represented: central-loop (**A**) and single-loop (**B**). In charts (**C**,**D**), data collected at station S10 (see Figure 2) using central-loop (**C**) and single-loop (**D**) devices are also shown.

It should be noticed that the shape of the curve, with changes in the tendency of increase or decrease in resistivity, allows detecting lithological changes. However, absolute values should not be taken into account: the dolomites offer average resistivity values in the 200–400 Ωm range (Figure 5), that can be lower than 100 Ωm in some stations (S1, S8, S9, and S10) and greater than 10^5 Ωm in others (S4). On the other hand, despite the overall similarity in the models of Figure 5A,B, a significant difference must be highlighted: the slope change of the decay curve of the induced potential. In the case of the central-loop device, the slope of the curve is lower, reaching depths of approximately 600 m with induced potential data on the order of 10 µV. With much lower voltage values, the curve continues to show a highly stable trend. However, using the single-loop device, the slope of the potential decay curve is greater. Thus, from approximately 150 m depth, the induced voltage values would be less than 10 μV.

The same comparison is performed at station S10 (see its location in Figure 2). Again, the good fit of the curves stands out, having an RMS error of 0.6% with a central loop ('qcenter_8' in Figure 5C) and 1.6% with a single loop ('q_2' in Figure 5D). With the central-loop device, three units are differentiated in depth: the shallowest (around 100 m depth) is characterized by a drop in resistivity values, although the local rises appearance would be associated to Messinian marls with sandstones interbedded. Between 100 and 300 m, conductive facies appear associated to Late Tortonian marls. Under these facies, the increase in resistivity values is associated to Jurassic dolomite aquifer. This interpretation also agrees with the information from direct data obtained in nearby outcrops (Figure 2B) and with the information provided by the seismic profile of Figure 3B.

In the case of the central-loop device, the potential curve reaches approximately a depth of 510 m with stable values (at approximately 150 m, the voltages are approximately 10 μV). In contrast, using the single-loop device, it offers values above 10 μV only up to 65 m, and the curve is no longer stable at a depth of 250. Therefore, at this station, the Jurassic aquifer is not reached with this configuration. From all this, it can be deduced that the central-loop device performs better than the single-loop device for deep aquifer research (Figures 5 and 6).

For the rest of the stations (S2 to S9), measurements with both configurations were made. For brevity, only the curves obtained by the central loop are shown (Figure 6). In all of them, the position of the Jurassic aquifer is detected at variable depths. In Figure 3A, the position of each of the TDEM stations on the seismic profile is shown. This allows us to compare the depth of Jurassic dolomites obtained by seismic (Figure 3) and electromagnetic methods (Figures 5 and 6).

Figure 7A shows the resistivity section obtained with the electromagnetic method, resulting from the joint inversion of data from the ten measurement stations through the ZondTem2D software. At the northern end of the section, the method is able to detect both the Jurassic dolomites and the contact with the underlying Paleozoic basement. In the southern sector, the Jurassic unit is also detected by a change in the trend of conductive to resistive materials. The method fits poorly in the central sector of the profile, where is unable to detect the decrease in resistivity values associated to the Triassic facies at deeper levels. The seismic reflection profile and the resistivity profile obtained by the TDEM method are represented together in Figure 7B. A good correlation can be observed between the two techniques for modeling the Jurassic unit. For the upper 200 m, at the northern end of the section, the TDEM resistivity profile and the profile obtained by the electrical resistivity tomography (ERT) method are also compared (Figure 7C). The superposition of the ERT and TDEM profiles allows deducing a good correlation in the lateral and vertical variations of the resistivity values obtained in both techniques.

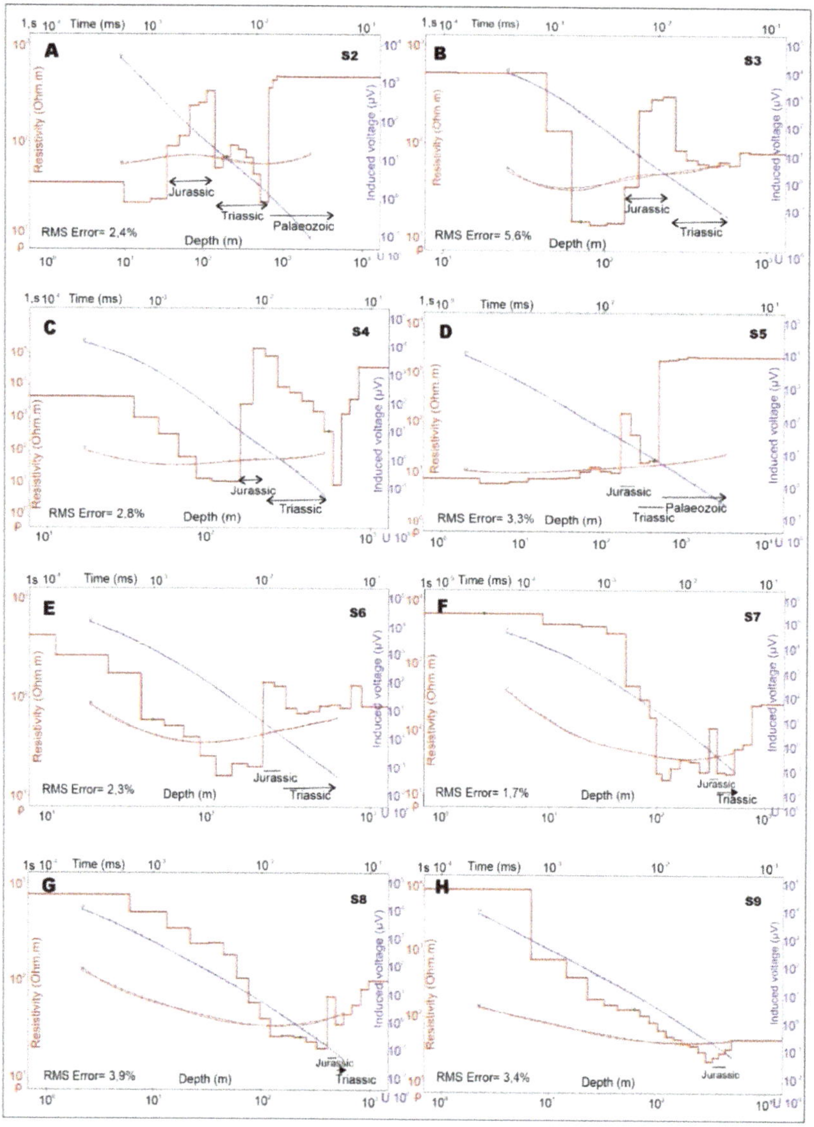

Figure 6. Representation of the induced voltage curves (blue line) and apparent resistivity (red line) as a function of time at stations S2 (**A**), S3 (**B**), S4 (**C**), S5 (**D**), S6 (**E**), S7 (**F**), S8 (**G**), and S9 (**H**) using a central-loop device. For each of them, the fit between the field line and the line of the ZONDTEM1D software model has been represented. The straight red lines represent the resistivity as a function of the depth, interpreted after data inversion. A good fit is deduced, with RMS error value ranges between 1.7% and 5.6%.

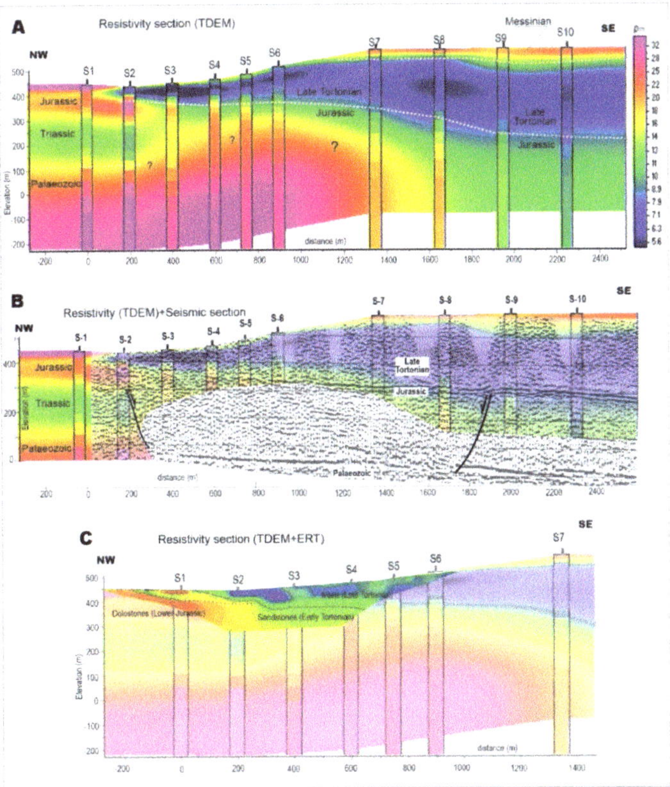

Figure 7. 2D section showing resistivity distribution as a result of smooth-model inversion implemented in the program ZondTEM2D (**A**). Seismic reflection profile and TDEM section correlation (**B**). Correlation of seismic reflection and electrical resistivity tomography (**C**).

5. Conclusions

In the seismic reflection profiles, abrupt changes in the amplitude of the wave associated to the upper and lower boundaries of the Jurassic dolomite aquifer are observed, making it easy to identify thickness variations at depth. Due to attenuation, such sudden changes in amplitude cannot be discerned in places where the aquifer is very shallow. At depths not exceeding 200 m, variations in resistivity values within ERT profiles allow identifying this aquifer level clearly. Simultaneously, the technique allows deductions of the lateral changes in facies, the thickness of the different units, structural dips, and the presence of faults.

The TDEM technique is versatile and can be used with different amplitudes, voltages, times, devices, and loop sizes, depending on the target penetration depths at which the research is scheduled. In our study, to reach more than 300 m depth, low voltages, high magnetic moments (increasing the amplitudes and the area of the loop), and central-loop-type devices were used. The thickness of the layer can be determined using TDEM curves, but not its real resistivity. In general, there is a good fit between seismic, electrical, and electromagnetic data.

Author Contributions: Conceptualization and investigation, J.R., J.M., R.M., S.S., and K.M.; software, V.T. and A.K.; writing and review, J.R., S.S., and M.C.H. All authors have read and agreed the published version of the manuscript.

Funding: This research received no external funding.

Acknowledgments: Careful reviews by R.M. Joeckel and by anonymous referees substantially improved the manuscript.

Conflicts of Interest: The authors declare no conflict of interest.

References

1. Griffiths, D.H.; Barker, R.D. Two-dimensional resistivity imaging and modelling in areas of complex geology. *J. Appl. Geophys.* **1993**, *29*, 211–226. [CrossRef]
2. Loke, M.H.; Barker, R.D. Rapid least-squares inversion of apparent resistivity pseudosections by a quasi-Newton method. *Geophys. Prospect.* **1996**, *44*, 131–152. [CrossRef]
3. André, F.; van Leeuwen, C.; Saussez, S.; Van Durmen, R.; Bogaert, P.; Moghadas, D.; de Resseguier, L.; Delvaux, B.; Vereecken, H.; Lambot, S. High-resolution imaging of a vineyard in south of France using ground-penetrating radar, electromagnetic induction and electrical resistivity tomography. *J. Appl. Geophys.* **2012**, *78*, 113–122. [CrossRef]
4. Binley, A.; Hubbard, S.S.; Huisman, J.A.; Revil, A.; Robinson, D.A.; Singha, K.; Slater, L.D. The emergence of hydrogeophysics for improved understanding of subsurface processes over multiple scales. *Water Resour. Res.* **2015**, *51*, 3837–3866. [CrossRef]
5. Dahlin, T.; Jens, E.; Dowen, R.; Mangeya, P.; Auken, E. Geophysical and hydrogeologic investigation of groundwater in the Karoo stratigraphic sequence at Sawmills in northern Matabeleland, Zimbabwe: A case history. *Hydrogeol. J.* **1999**, *15*, 945–960.
6. Yechieli, Y.; Kafri, U.; Goldman, M.; Voss, C.I. Factor controlling of configuration of the fresh-saline water interface in the Dead Sea coastal aquifer: Synthesis of TDEM surveys and numerical groundwater modeling. *Hydrogeol. J.* **2001**, *9*, 367–377.
7. Boucher, M.; Favreau, G.; Descloitres, M.; Vouillamoz, J.M.; Massuel, S.; Nazoumou, Y.; Legchenko, A. Contribution of geophysical surveys to groundwater modelling of a porous aquifer in semiarid Niger: An overview. *Comptes Rendus Geosci.* **2009**, *341*, 800–809. [CrossRef]
8. Rey, J.; Redondo, L.; Hidalgo, M.C. Interés hidrogeológico de las dolomías del Liásico de la Cobertera Tabular de la Meseta (norte de Úbeda, prov. de Jaén). *Rev. Soc. Geol. Esp.* **1998**, *11*, 213–221.
9. Gonzalez-Ramón, A.; Heredia, J.; Rodríguez-Arévalo, J.; Manzano, M.; Ortega, L.; Muñoz de la Varga, D.; Moreno, J.A.; Díaz Teijeiro, M.F. Evolución temporal de las características físico-químicas e isotópicas en el agua subterránea de los acuíferos de la Loma de Úbeda (sur de España). *Asam. Hisp. Port. Geod. Geofís. Donostia Proc.* **2012**, *7*, 427–434.
10. I.T.G.E. *Atlas Hidrogeológico de la Provincia de Jaén*; ITGE-Diputación Provincial de Jaén: Jaén, España, 2012.
11. Fernández, J.; Dabrio, C. Fluvial Architecture of the Buntsandstein-facies redbeds in the Middle to Upper Triassic (Ladinian-Norian) of the southeastern edge of the Iberian Meseta (Southern Spain). In *Aspects of Fluvial Sedimentation in the Lower Triassic Buntsandstein of Europe*; Lecture Notes in Earth Sciences; Mader, D., Ed.; Springer: Berlin, Germany, 1985; Volume 4, pp. 411–435.
12. Vera, J.A.; López-Garrido, A.C. Sobre las facies detríticas rojas (red beds) del borde sureste de la Meseta. *Cuad. Geol.* **1971**, *2*, 147–155.
13. Azcárate, J.E. *Mapa Geológico y Memoria Explicativa de la Hoja 906 (Úbeda) del Mapa Geológico de España, Escala 1:50.000*; I.G.M.E.: Madrid, Spain, 1977.
14. Rey, J.; Redondo, L.; Aguado, R. Relleno, durante el Tortoniense inferior, de paleocubetas en las proximidades de Úbeda (provincia de Jaén). *Bol. Geol. Min.* **1995**, *106*, 215–218.
15. Pedrera, A.; Ruiz-Constán, A.; Marín-Lechado, C.; Galindo-Zaldívar, J.; González, A.; Peláez, J.A. Seismic transpressive basement faults and monocline development in a foreland basin (Eastern Guadalquivir, SE Spain). *Tectonics* **2013**, *32*, 1571–1586. [CrossRef]
16. Sánchez Gómez, M.; Peláez, J.A.; García Tortosa, F.J.; Pérez Valera, F.; Sanz de Galdeano, C. La serie sísmica de Torreperogil (Jaén, Cuenca del Guadalquivir oriental): Evidencias de deformación tectónica en el área epicentral. *Rev. Soc. Geol. Esp.* **2014**, *27*, 301–318.
17. Perconing, E. Sobre la edad de la Transgresión del Terciario marino en el borde meridional de la Meseta. In *Congr. Hispano-Luso-Americano de Geol. Económica*; Ibérica-Tarragona: Madrid, España, 1971; Section 1; pp. 309–319.

18. Hidalgo, M.C.; Rey, J.; Redondo, L. Origen de la surgencia del Balneario de San Andrés (Provincia de Jaén). *Geogaceta* **1998**, *24*, 179–182.
19. González-Ramón, A.; Rodríguez-Arévalo, J.; Martos-Rosillo, S.; Gollonet, J. Hydrogeological research on intensively exploited deep aquifers in the 'Loma de Úbeda' area (Jaén, southern Spain). *Hydrogeol. J.* **2013**, *21*, 887–903. [CrossRef]
20. Store, H.; Storz, W.; Jacobs, F. Electrical resistivity tomography to investigate geological structures of earth's upper crust. *Geophys. Prospect.* **2000**, *48*, 455–471. [CrossRef]
21. Telford, W.M.; Geldart, L.P.; Sheriff, R.E. *Applied Geophysics*; Cambridge University Press: Cambridge/London, UK, 1990; 770p.
22. Sasaki, Y. Resolution of resistivity tomography inferred from numerical simulation. *Geophys. Prospect.* **1992**, *40*, 453–464. [CrossRef]
23. Maillet, G.M.; Rizzo, E.; Revil, A.; Vella, C. High resolution electrical resistivity tomography (ERT) in a transition zone environment: Application for detailed internal architecture and infilling processes study of a Rhône River paleo-channel. *Mar. Geophys. Res.* **2005**, *26*, 317–328. [CrossRef]
24. Martínez, J.; Benavente, J.; García-Aróstegui, J.L.; Hidalgo, M.C.; Rey, J. Contribution of electrical resistivity tomography to the study of detrital aquifers affected by seawater intrusión-extrusion effects: The river Vélez delta (Vélez-Málaga, southern Spain). *Eng. Geol.* **2009**, *108*, 161–168. [CrossRef]
25. Rey, J.; Martínez, J.; Hidalgo, C. Investigating fluvial features with electrical resistivity imaging and ground-penetrating radar: The Guadalquivir River terrace (Jaen, Southern Spain). *Sediment. Geol.* **2013**, *295*, 27–37. [CrossRef]
26. Goldman, M.; Rabinovich, B.; Rabinovich, M.; Gilad, D.; Gev, I.; Shirov, M. Application of the integrated NMR-TDEM method in groundwater exploration in Israel. *J. Appl. Geophys.* **1994**, *31*, 27–52. [CrossRef]
27. Auken, E.; Jørgensen, F.; Sørensen, K.I. Large-scale TEM investigation for groundwater. *Explor. Geophys.* **2003**, *34*, 188–194. [CrossRef]
28. Martínez-Pagán, P.; Faz Cano, A.; Aracil, E.; Arocena, J.M. Electrical resistivity imaging revealed the spatial properties of mine tailing ponds in the Sierra Minera of Southeast Spain. *J. Environ. Eng. Geophys.* **2009**, *14*, 63–76. [CrossRef]
29. Cortada, U.; Martínez, J.; Rey, J.; Hidalgo, M.C. Assessment of tailings pond seals using geophysical and hydrochemical techniques. *Eng. Geol.* **2017**, *223*, 59–70. [CrossRef]
30. Dahlin, T.; Zhou, B. A numerical comparison of 2D resistivity imaging with 10 electrode arrays. *Geophys. Prospect.* **2004**, *52*, 379–398. [CrossRef]
31. Loke, M.H. Tutorial: 2-D and 3-D Electrical Imaging Surveys; Revision date: 5 November 2014. Available online: www.geotomosoft.com (accessed on 20 October 2019).
32. Loke, M.H.; Dahlin, T. A comparison of the Gauss-Newton and quasi-Newton methods in resistivity imaging inversion. *J. Appl. Geophys.* **2002**, *49*, 149–162. [CrossRef]
33. Nabighian, N.M. *Electromagnetic Methods in Applied Geophysics-Theory*; Society of Exploration Geophysics: Tulsa, OK, USA, 1988; 971p.
34. Spies, B.R. Depth of investigation in electromagnetic sounding methods. *Geophysics* **1989**, *54*, 872–888. [CrossRef]
35. Fitterman, D.V.; Stewart, M.T. Transient electromagnetic sounding for groundwater. *Geophysics* **1986**, *51*, 995–1005. [CrossRef]
36. Gómez-Ortiz, D.; Fernández-Remolar, D.C.; Granda, A.; Quesada, C.; Granda, T.; Prieto-Ballesteros, O.; Molina, A.; Amils, R. Identification of the subsurface bodies responsible for acidity in Rio Tinto source water; Spain. *Earth Planet. Sci. Lett.* **2014**, *391*, 36–41. [CrossRef]
37. Xue, G.Q.; Qin, K.Z.; Li, X.; Li, G.M.; Qi, Z.P.; Zhou, N.N. Discovery of a Large-scale Porphyry Molybdenum Deposit in Tibet through a Modified TEM Exploration Method. *J. Environ. Eng. Geophys.* **2012**, *17*, 19–25. [CrossRef]
38. Granda Sanz, A.; Granda París, T.; Pons, J.M.; Videira, J.C. El descubrimiento del Yacimiento de la Magdalena. Protagonismo de los métodos geofísicos en la exploración de yacimientos tipo sulfuros masivos vulcanogénicos (vms) profundos en la faja pirítica ibérica. *Bol. Geol. Min.* **2019**, *130*, 213–230. [CrossRef]
39. Danielsen, J.E.; Auken, E.; Jorgensen, F.; Sondergaard, V.; Sorensen, K. The application of the transient electromagnetic method in hydrogeophysical survey. *J. Appl. Geophys.* **2003**, *53*, 181–198. [CrossRef]

40. Dennis, Z.R.; Cull, J.P. transient electromagnetic survey for the measurement of near-surface electrical anisotropy. *J. Appl. Geophys.* **2012**, *76*, 64–73. [CrossRef]
41. Sridhar, M.; Markandeyulu, A.; Chaturvedi, A.K. Mapping subtrappean sediments and delineating structure with the aid of heliborne time domain electromagnetics: Case study from Kaladgi Basin, Karnataka. *J. Appl. Geophys.* **2017**, *136*, 9–18. [CrossRef]
42. Hallbauer-Zadorozhnaya, V.Y.; Stettler, E. Time Domain Electromagnetic Sounding to delineare hydrocarbon Contamination of ground water. In *Symposium on the Application of Geophysics to Engineering and Environmental Problems*; European Association of Geoscientists & Engineers: Fort Worth, TX, USA, 2009; pp. 241–251.
43. Hu, Y.; Zeng, Z.; Chen, X.; Zhao, X. Application of Transient Electromagnetic Method in saturation region detection in tailings dam. In *7th International Conference on Environmental and Engineering Geophysics & Summit Forum of Chinese Academy of Engineering on Engineering Science and Technology*; Atlantis Press: Beijing, China, 2016; pp. 318–321.
44. Kafri, U.; Goldman, M. Are the lower subaquifers of the Mediterranean coastal aquifer of Israel blocked to seawater intrusion? Results of a TDEM (time domain electromagnetic) study. *Isr. J. Earth Sci.* **2006**, *55*, 55–68. [CrossRef]
45. Levi, E.; Goldman, M.; Hadad, A.; Gvirtzman, H. Spatial delineation of groundwater salinity using deep time domain electromagnetic geophysical measurements: A feasibility study. *Water Resour. Res.* **2008**, *44*, 1–14. [CrossRef]
46. Christiansen, A.V.; Auken, E.; Sorensen, K. The transient electromagnetic method. In *Groundwater Geophysics. A Tool for Hydrogeology*; Kirsch, R., Ed.; Springer: Berlin, Germany, 2009; pp. 179–224.

applied sciences

MDPI

Article

Characterization of a Shallow Coastal Aquifer in the Framework of a Subsurface Storage and Soil Aquifer Treatment Project Using Electrical Resistivity Tomography (Port de la Selva, Spain)

Alex Sendrós [1,2,*], Aritz Urruela [1], Mahjoub Himi [1], Carlos Alonso [3], Raúl Lovera [1,2], Josefina C. Tapias [2,4], Luis Rivero [1], Ruben Garcia-Artigas [2] and Albert Casas [1,2]

[1] Department of Mineralogy, Petrology and Applied Geology, Universitat de Barcelona, 08028 Barcelona, Spain; aritz.urruela@ub.edu (A.U.); himi@ub.edu (M.H.); rlovera@ub.edu (R.L.); lrivero@ub.edu (L.R.); albert.casas@ub.edu (A.C.)
[2] Water Research Institute, Universitat de Barcelona, 08001 Barcelona, Spain; jtapias@ub.edu (J.C.T.); r.garcia@ub.edu (R.G.-A.)
[3] Seequent, Marlow SL7 1DB, UK; carlos.alonso@seequent.com
[4] Department of Biology, Health and Environment, Universitat de Barcelona, 08028 Barcelona, Spain
* Correspondence: alex.sendros@ub.edu

Citation: Sendrós, A.; Urruela, A.; Himi, M.; Alonso, C.; Lovera, R.; Tapias, J.C.; Rivero, L.; Garcia-Artigas, R.; Casas, A. Characterization of a Shallow Coastal Aquifer in the Framework of a Subsurface Storage and Soil Aquifer Treatment Project Using Electrical Resistivity Tomography (Port de la Selva, Spain). *Appl. Sci.* **2021**, *11*, 2448. https://doi.org/10.3390/app11062448

Academic Editor: Francisco Javier Alcalá

Received: 5 February 2021
Accepted: 6 March 2021
Published: 10 March 2021

Abstract: Water percolation through infiltration ponds is creating significant synergies for the broad adoption of water reuse as an additional non-conventional water supply. Despite the apparent simplicity of the soil aquifer treatment (SAT) approaches, the complexity of site-specific hydrogeological conditions and the processes occurring at various scales require an exhaustive understanding of the system's response. The non-saturated zone and underlying aquifers cannot be considered as a black box, nor accept its characterization from few boreholes not well distributed over the area to be investigated. Electrical resistivity tomography (ERT) is a non-invasive technology, highly responsive to geological heterogeneities that has demonstrated useful to provide the detailed subsurface information required for groundwater modeling. The relationships between the electrical resistivity of the alluvial sediments and the bedrock and the difference in salinity of groundwater highlight the potential of geophysical methods over other more costly subsurface exploration techniques. The results of our research show that ERT coupled with implicit modeling tools provides information that can significantly help to identify aquifer geometry and characterize the saltwater intrusion of shallow alluvial aquifers. The proposed approaches could improve the reliability of groundwater models and the commitment of stakeholders to the benefits of SAT procedures.

Keywords: aquifer geometry; electrical resistivity tomography; managed aquifer recharge; saltwater intrusion; soil aquifer treatment; sustainable development goal 6

1. Introduction

Surface water resources in coastal areas are often scarce and groundwater plays a pivotal role in managing the complex issue of water supply [1]. Urbanization and climate change are causing several side effects for the sustainable management of groundwater resources and in parallel, the volume of wastewater rises as the population increases [2–4]. Moreover, coastal area development is often intensive and subject to salinity problems [5]. The intensive extraction of groundwater from coastal aquifers reduces freshwater outflow to the sea and creates local water table depression, causing seawater to migrate inland and rising toward the wells [6]. This phenomenon is called seawater intrusion. It is the consequence of meeting the increasingly urban, tourism, industrial, and agricultural demands and adds stress to groundwater bodies and dependent ecosystems [7].

These effects are especially visible in small unconfined aquifers because of their greater exposure to human impacts and smaller storage capacity, which may limit a regular water provision to ecosystems during drier periods. They will require efficient adaptations at the scale of the municipality or group of municipalities to supply the demand, to deal with the expected more irregular periods of rain, a decrease of aquifer recharge, and to manage wastewater properly. The most obvious is the provision of an alternative water resource and the second is the reduction of environmental impacts by reducing or eliminating wastewater disposal, which results in the preservation of ecosystems downstream [8].

To address these water resilience challenges, managed aquifer recharge schemes using recycled water have been utilized during the last few decades [9]. Water reuse is thus a means for increasing water resources and the subsoil possesses key advantages to help water operators. Aquifers offer a reservoir—where there is no room for surface storage—and uses the capacity for inter-seasonal and inter-year storage that natural systems provide [10,11]. Systems like direct well injection or infiltration with either ponds, basins, or shallow buried trenches have been successfully implemented in small to large scale projects [12–14]. Furthermore, in coastal areas affected by saltwater intrusion, both systems may endure a particularly efficient way to tackle contamination because the water infiltration acts as a barrier to saline encroachment and, at the same time, allows the recovery of salinized aquifers [15].

Well injection offers advantages such as no evaporative losses, algae, or mosquitoes, and no loss of prime land. Alternatively, surface infiltration offers a potentially low-cost way to reclaim wastewater during the migration of recycled water through the vadose zone, which can result in improved water quality via biological, chemical, and physical processes before reaching the water table [16].

Schemes using the favorable characteristics of soil, subsoil, and aquifer for further (advanced) treatment of the infiltrated water are called SAT (Soil Aquifer Treatment) and usually have severe restrictions in terms of water quality used for the procedure [17]. Water percolation through infiltration basins is currently creating significant synergies for the broad adoption of water reuse as an alternate water supply and could foster sustainable water consumption in a transition to a circular economy.

Despite the apparent simplicity of the surface SAT approaches, the complexity of site-specific hydrogeological conditions and the processes occurring at various scales, combined with different objectives, requires an excellent understanding of the system's response [18,19]. A groundwater model is required according to published guidelines for managed aquifer recharge and water recycling [20].

Groundwater modeling tools involve three general stages for aquifer conceptualization: (1) geometry definition, (2) acquisition of hydraulics data, and (3) evaluation of water balance components [21]. The paper's general objective is advancing the first stage, i.e., aquifer geometry definition, and qualifying some transient groundwater features subordinately (saltwater intrusion). Classical hydrogeological approaches such as using lithological logs, aquifer water sampling, and measuring water tables require the existence of available detailed drilling reports of wells and piezometers and/or drill and equipping new ones. Besides, such logs provide only punctual assessments of the aquifer geometry and hydraulic data.

In flat coastal areas, near-surface geophysical techniques have been widely used in groundwater research to acquire this basic information. Geophysical techniques are high-resolution tools that provide information on the spatial distribution of the physical parameters of the subsoil. These techniques are non-invasive and become useful when lithological data are sparse or not able to provide detailed subsurface information required for groundwater modeling [21,22]. The advantage of the surface geophysical methods, compared with borehole methods, is that they allow denser and faster data coverage at a much lower cost [23], reduce risks to interconnect different water bodies during drilling operations, and can provide a way to improve the subsurface characterization from a set of boreholes.

The most widely used method for hydrological applications and shallow aquifers characterization is probably the direct-current (DC) resistivity, which is highly responsive to detailed subsurface electrical conductivity changes related to geological and hydrological heterogeneities [24]. Electrical resistivity (also called DC resistivity) methods measure the apparent electrical resistivity of the formation and have been widely used to delineate the geometry of shallow aquifers [25–27] and saltwater intrusion [28–32]. However, ERT has rarely been used in SAT projects framework and it could define better the boundary conditions in its hydrogeological models, which is important for understanding and predicting subsurface flow and transport.

The specific aim of our research was to develop a fast and non-invasive methodology to improve the characterization of the Port de la Selva shallow coastal aquifer and delineate the saltwater intrusion. To this end, we propose an approach using electrical resistivity tomography and implicit 3D modeling tools [33] to define the aquifer geometry and the saltwater intrusion. Particularly, a 2D ERT campaign was conducted to assess the thickness and lateral extent of the alluvial formation below the profiles acquired and to gather data for subsequent 3D modeling. Placing ERT data in the same tool framework allowed to delimit a saltwater intrusion phenomenon. The information from the geological map and research boreholes was also integrated to define boundary model conditions and to obtain the aquifer geometry of the whole studied area respectively.

2. Study Area

2.1. Geographical, Climate, and Water Management Overview

The study area is located on the northeast coast of the Iberian Peninsula, at Port de la Selva valley. The valley covers an area of 10.5 km^2, and mainly extends over the north and east slope of the Rodas Mountains (with a peak elevation of 670 m above sea level) until they reach the Mediterranean Sea.

The area is characterized by a Mediterranean climate with a warm thermal regime in summer and moderate cold in winter. Average monthly temperatures have varied from 8.8 °C in February to 23.7 °C in July over the period 2007–2016. The annual rainfall is moderate, with an average of 570 mm/year, but below 450 mm/year in dry years. Rainwater is distributed irregularly during the year. Maximum rain values are recorded in the October–November period (more than 160 mm on average) and the driest month is July (below 20 mm), coinciding with the highest potential evapotranspiration values [34].

The rainfall is assumed to be mainly incorporated into surface run-off (60%) while evapotranspiration (25%) and aquifer infiltration (15% or equivalent to 2000 m^3/day) completed the water balance of the Port de la Selva basin [35].

Similar to other Mediterranean coastal areas, Port de la Selva municipality has experienced extensive urbanization and an associated shift away from an agricultural and fishing economy to one governed by the growth of tourism from the 1950s to the present (Figure 1). Water demands and wastewater effluents in the area have therefore greatly increased—due to the rise of water consumption for recreational uses and tourists—especially in summer, when its population reaches 10,000 inhabitants compared to 1000 inhabitants during winter or non-tourist season. The average annual supply water abstraction in El Port de la Selva basin is about 400,000 m^3, while the average annual volume of treated wastewater is in the range of 220,000 m^3. Both abstraction and wastewater volumes have large fluctuations between summer and winter [36]. Furthermore, chloride concentrations in water abstracted from the municipal well were recurringly above the regional drinking water limit of 250 mg/L [37]—during the autumn months—probably triggered by saltwater intrusion after summer's high water demand period (Figure 2).

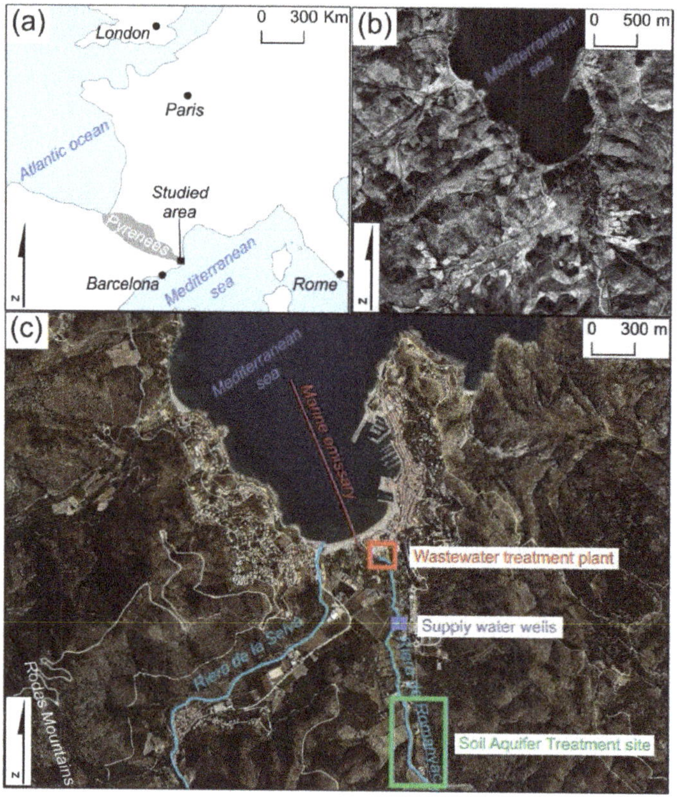

Figure 1. (**a**) Location sketch of the studied area; (**b**) landscape image of Port de la Selva (1956) showing the extensive agricultural use of soil in the 1950s compared to the current urbanization setting; (**c**) location of main hydrological features and water management facilities over current landscape image of Port de la Selva.

To increase the security of water supplies and to improve the water quality, Port de la Selva's town council have constructed an SAT—pioneering in Europe in the usage of recycled wastewater for recharging an aquifer dedicated to human consumption—and joined the European Commission DEMOWARE project. The DEMOWARE project involves the public perception in water reuse, artificial recharge of water into aquifers and disseminate investigations related to wastewater recycling.

The innovation and significance of the project were to demonstrate the SAT commissioning without the need for advanced treatment such as ozonation or reverse osmosis, relying on the natural treatment capacity of the soil and aquifer only. The used waters of the municipality of Port de la Selva are connected to the sanitation system consisting of a wastewater treatment plant (WWTP) with biological flocculation, tertiary double-filtration, and UV disinfection treatments. Most of the water entering the WWTP receives biological treatment with nitrogen removal (approximately 70%) and the effluent is discharged into the Romanyac stream (during the period October to May) or directly into the sea through the marine emissary (the months of June, July, August, and September). The Romanyac stream outlets to the mid part of the beach and the point of discharge of the effluent of the WWTP is about 300 m from the coastline. DEMOWARE project proposes that wastewater, after receiving an appropriate tertiary treatment could be reused for aquifer recharge SAT system. For this purpose, the reclaimed wastewater would be pumped to the SAT infiltration ponds where the subsoil acts as a low-cost/low-energy filtration/disinfection scheme including seasonal storage. To close the circle, artificially recharged water—diluted into the natural groundwater flow—is pumped on wells placed on the same alluvial aquifer about 1 km downstream.

Risk analysis results derived from the DEMOWARE project showed that the system was efficient for reducing mononuclear phagocyte system (MPS) bacteria content below the World Health Organization (WHO) and Spanish regulations and removed after 50 m of flow through the subsoil. Nevertheless, the start-up of the recharging system is still dependent on obtaining authorization from the appropriate health care authorities.

Handling a flow and transport model, the migration of the water infiltrated in the aquifer from the recharge ponds was simulated by Amphos21 to analyze the sensitivity of travel times and dilution factors related to different scenarios [38]. Moreover, the numerical model was capable of simulating aquifer response to rainfall events and pumping in water supply wells and reproduces the observed hydraulic heads with reasonable accuracy [35]. The model has been constrained from 12 monitoring wells. Nevertheless, it is expected that geophysical data could improve the reliability of the groundwater models representing the heterogeneity of the shallow aquifer.

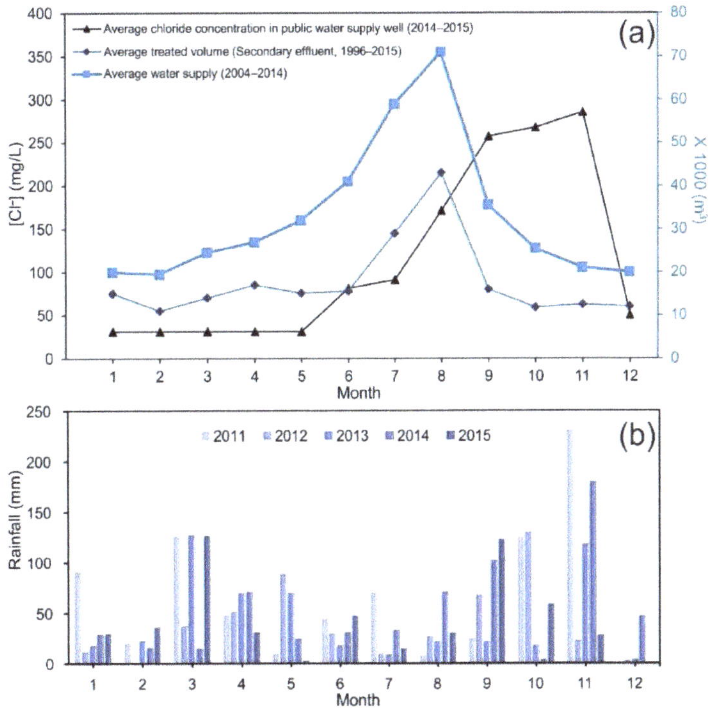

Figure 2. (**a**) Seasonal fluctuations of treated and abstracted water volumes and measured chloride concentration in Port de la Selva's public water supply well [39]. The graph shows the correlation between water demand and wastewater generation and the deferred effect between maximum exploitation and salinization of the water supply well. The dashed line represents the chloride concentration limit defined by the Spanish sanitary guidelines for the quality of water for human consumption [37]; (**b**) rainfall for the period 2011–2016. The summer season corresponds to the dry months of the year and the highest water demand.

2.2. Geological and Hydrogeological Setting

The study area belongs to the easternmost outcrop of the Hercynian basement of the Axial Zone of the Pyrenees. The Axial Zone consists of an elongated structure along the Pyrenean chain. In this Axial Zone, together with the North Pyrenean massifs, rocks from the Hercynian basement outcrop due to both Hercynian and Alpine tectonic uplift [40]. Locally, two geological units can be differentiated in the study area: (1) Palaeozoic basement includes (i) schists, slates, and phyllites and (ii) amphybolites all affected by the Hercynian regional metamorphism, and (2) Pos-Hercynian sedimentary cover includes (i) Upper Pleistocene deposits of colluvial and aeolian origin and (ii) Holocene alluvial deposits [41].

In hydrogeological terms, lithological units can be classified also into two groups according to their permeability [42]: (1) Cambro–Ordivician schists and phyllites of low-hydraulic conductivity forming the impervious basement of the alluvial aquifer; (2) quaternary deposits of gravel and sand with variable silt content (water stream alluvial deposits) and sand and clay with pebbles of high hydraulic conductivity acting as an unconfined aquifer (Figure 3). The contact between the two formations was identified at depths ranging from eight to 14 m in the Port de la Selva site, but there is a lack of additional data in the overall studied area ($1 \cdot 10^8$ m^2) as the exploration boreholes (Figure 4) are mainly clustered around the SAT facilities ($3 \cdot 10^5$ m^2).

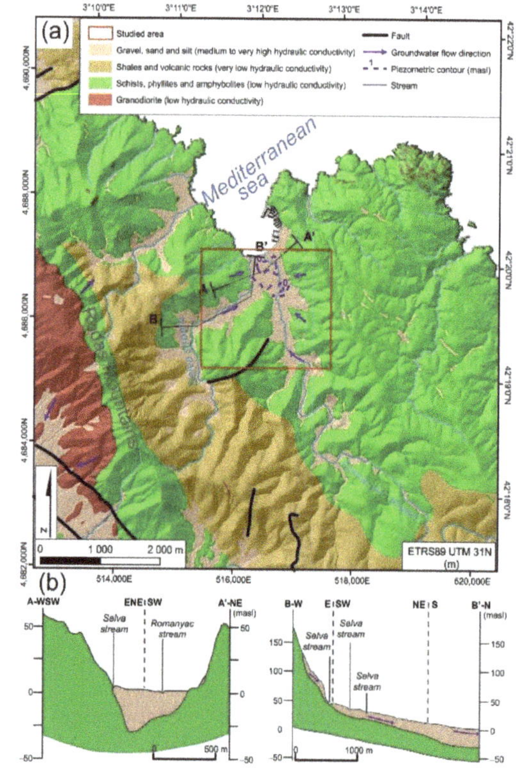

Figure 3. (**a**) Hydrogeological setting of Port de la Selva; (**b**) simplified geological cross-sections of Port de la Selva coastal area. Modified from References [42,43].

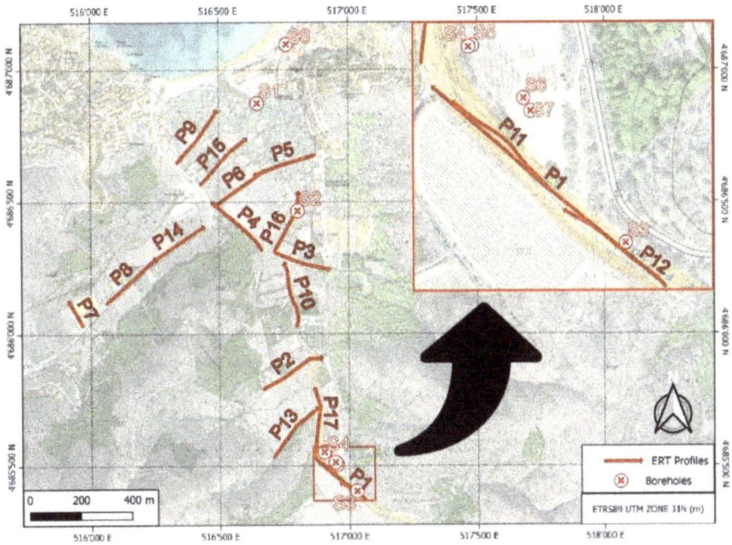

Figure 4. Location of the electrical resistivity tomography profiles and boreholes (Port de la Selva).

Groundwater flow through the metamorphic substrate is controlled by faults and joints (secondary porosity) of the otherwise impermeable rock. For this reason, the flow and water storage capacity of this considered secondary aquifer is limited to discrete planes. Then, for practical purposes, the water resources of the secondary aquifer are irrelevant, except for the most important faults.

The alluvial aquifer is located at the bottom of the valley, over the metamorphic bedrock. It has a surface of 0.66 km^2 and could reach 30 m in thickness [43]. The hydraulic conductivity values range from 50 to 250 m/day. The relatively small total volume and high topographic gradient at the top of the basin mean that it does not have a very high storage capacity [35]. It is assessed as moderate to high on groundwater pollution vulnerability [43]. Recharge to alluvial aquifer comes from direct rainfall and runoff from intermittent water streams of Riera de la Selva and Riera de Romanyac and water outflows are due to pumping and discharge to the Mediterranean Sea. Water supply wells operate in this section of the alluvial aquifer with estimated groundwater withdrawn of 500–600 m^3/day in winter and 2500 m^3/day in summer.

3. Materials and Methods

3.1. Electrical Resistivity Tomography

In order to assess the thickness and lateral extent of the alluvial aquifer, an electrical resistivity survey (ERT) was conducted. The resistivity method is based on measuring the potentials between one electrode pair while transmitting DC between another electrode pair. The depth of penetration is proportional to the separation between the electrodes, and by varying the electrode separation, information is provided on the subsurface stratification. This geophysical technique can be considered the modern evolution of the classical geoelectrical methods, as the vertical electrical sounding. In fact, the physical principle is the same, but in this case, instead of using only four electrodes (two for energizing and two for measuring the potential generated), multiple electrodes that change function automatically are fixed in the soil surface. All possible combinations of electrode pairs are considered, resulting in a dataset of apparent resistivities at the so-called pseudo-depth at different locations. The large amount of data produced by multielectrode systems requires automated data handling and processing [44].

In the present study, the Wenner–Schlumberger array was used. The apparent resistivity for the array is given by $\rho_a = \pi n(n + 1) \cdot a \cdot R$, where R represents the resistance, a is the spacing between the potential electrodes, and n is the ratio of the distances between the current and potential electrodes [28]. The depth study range increases with increasing space between the current electrodes, whereas a shorter separation increases resolution [45].

The software for the inversion of the ERT data was RES2DInv [46]. The subsurface is divided into cells of fixed dimensions and the inversion procedure is based on the smoothness-constrained least-squares method. The resistivities are adjusted iteratively until a satisfactory agreement between the experimental data and the model responses is achieved, based on a nonlinear optimization technique by least-squares fitting [47]. During the inversion process, the root-mean-square value of the difference between experimental data and the updated model response is used as a criterion to assess the convergence.

In the present paper, the smooth constraint method was selected, after making a comparison with the robust method. The method assumes the subsurface consists of a few homogeneous regions with a smooth interface between them. Such an inversion scheme is the logical choice where the subsurface comprises units with smooth boundaries in order to determine both layer boundary locations and layer resistivities accurately. Indeed, it produces models by minimizing the absolute value of data misfit, making it more efficient in removing noise compared to other inversion methods [48].

The subsequent subsurface characterization using electrical conductivity or resistivity depends on several factors, such as soil water content, grain size distribution, porosity, and permeability. For instance, an air-filled void soil type will have higher geoelectrical resistivity values contrary to a water-filled void soil type [49].

Resistivity decreases with increasing salinity. A high-salinity pore fluid has a greater concentration of ions available for conduction. Besides, igneous and metamorphic rocks typically have high resistivity values while resistivity decreases as grain size particles decrease in unconsolidated sediments.

Lastly, the geological interpretation of the resistivity cross-section is performed incorporating, as far as possible, prior knowledge based on outcrops, supporting geophysical or borehole data, and any information gained from laboratory studies of the electrical resistivity of geological materials [50].

3.2. Electrical Resistivity Tomography Surveys

ERT data was acquired with a Syscal Pro resistivity meter (IRIS instruments, Orléans, France). The system features an internal switching board for 72 electrodes and an internal 250 W power source. The Wenner–Schulmberger array was chosen because it is properly sensitive to both horizontal and vertical structures and has a relatively good signal strength [51]. The configuration has high performance and stability in high electrical resistivity environments such as dry gravels and it is useful for the investigation of horizontal or slightly inclined layers that can present lateral facies changes and/or verticalized structures, as is the case of the studied site [52,53].

The design of geophysical surveys and the selection of the multi-electrode arrays was planned considering the length available for the acquisition, depth of investigation, the resolution required, and the expected structure derived from hydrogeological background knowledge. Lithological logs of boreholes drilled for DEMOWARE Project (borehole S3 to S7) and from a new building project (borehole S8) have been an invaluable support for our research (Figure 4).

The distribution of the 17 ERT acquired profiles was affected by the site physical barriers (buildings, fences, roads . . .) and the availability of space to extend the arrays along an almost straight line. Acquiring ERT using straight lines increases the efficiency of the survey as it is not necessary adapting on-site the array to a different geometry setting. The objective was to cover the study area with a representative network of the variability of electrical resistivity values with profiles distributed as homogeneously as possible in the area. As a result, six cross-sections were acquired perpendicular, seven transversal, and two obliques to the direction of the main water streams. Additionally, two detailed cross-sections (P11 and P12) were gathered near piezometers S3 to S7 and cross-section P1. Their purpose was to increase the resolution in the area with more geological data and correlate resistivity values and lithologic changes (Table 1).

Table 1. Main characteristics of the ERT profiles (Port de la Selva).

ID	Number of Profiles	Length (m)	Electrode Interval (m)	Orientation to Water Stream
P1 and P2	2	235	5	parallel and transverse
P3	1	188	4	oblique
P4, P5 and P6	3	235	5	parallel and transverse
P7	1	94	2	transverse
P8	1	235	5	parallel
P9 and P10	2	235	5	transverse and parallel
P11 and P12	2	94	2	parallel
P13 to P17	5	235	5	transverse, parallel and oblique

The 235 m long 2D ERT cross-sections allowed us to reach a research-depth close to 50 m—allowing to characterize the expected maximum aquifer thickness of 30 m—and a resolution of one point every five meters in both directions. The 94 m length 2D ERT detailed cross-sections allowed to reach a research-depth close to 20 m—the aquifer boundary was assumed at 10–15 m depth—and a resolution of two meters apart between geoelectrical values.

3.3. *Aquifer Geometry and Saltwater Intrusion*

The aquifer geometry is used to define boundary conditions in hydrogeological modeling and outlining saltwater intrusion to characterize transient groundwater features in coastal aquifers. We have used Leapforg Geo v 6.0. software [54] for deriving the aquifer geometry model from discrete variables such as lithological information from boreholes and ERT data and ERT data for deriving a saltwater intrusion model.

Geoelectrical electrical data was positioning first using a differential GRS1 GPS instrument (Topcon, Itabashi, Japan), and relative relief profiles of ERT cross-sections were converted into georeferenced elevation profiles using an earth digital elevation model provided by the Catalan Geographical and Geological Institute (ICGC). The elevation model has a 2 × 2 m resolution and its estimated absolute vertical accuracy corresponds to an average mean quadratic error of 0.15 m in flat and low vegetation areas.

Placing all this data in the same framework allowed to delimit the aquifer-aquiclude contact in the software. First, the ERT data contrasts—in the resistivity values—mark the boundary as 2D lines and provide great detail. Next, the 1D information from the boreholes reaching the substratum was incorporated. Finally, the geologic map data [41] were also integrated into the model to include the plan view contact of the two units as an additional boundary condition.

Leapfrog workflow is based on an implicit modeling method and on creating contact surfaces between different lithologies. Afterward, these surfaces are activated, and they "cut" finite volume into respective units. Adopting this approach of starting with a finite volume and using contact surfaces to "cut" it into units means that, inherently, there will be no void space or overlapping volumes in the geological model.

Leapfrog geo was also used to interpolate the ERT data. Leapfrog Geo uses FastRBF™, a mathematical algorithm developed from radial basis functions (RBF). FastRBF employs the numerical or categorical data and parameters supplied to derive any one of a number of variables to be modeled in 3D space. RBFs are a family of interpolation functions that were first introduced into the geological literature by Hardy [55] to interpolate scattered topographic data. RBF techniques have been considered good surface interpolators due to their attempt to honor raw data [56] and their ability to provide the smoothest surface of interpolation [57,58], which is ideally suited for geological modeling [59,60]. Implicit geologic modeling using RBFs is also comparable in quality to modeling using popular co-kriging approaches [61,62]. ERT data were interpolated using this method inside the boundaries of the geological model created in the previous steps.

Moreover, new information from borehole and resistivity data has been progressive, fast, and dynamically incorporated into the aquifer geometry and saltwater intrusion models using implicit modeling tools. Fast 3D geological modeling tools technological breakthrough has not been successful so far in traditionally oriented near-surface or aquifer characterization, except in the mining and oil exploration markets with budgets on many different scales [63–65].

With the advent of fast 3D interpolation methods [66], the construction of geological surfaces using volume functions such as RBFs is now a practical alternative to explicit modeling of surfaces. Unlike explicit modeling, surfaces contained in volume functions are not explicitly defined or digitized. Instead, the existence of surfaces in the volume function is implicit. Based on recent advances in fast scattered data interpolation methods, implicit modeling first defines a continuous three-dimensional function that describes the rock changes distribution. This volumetric function is interrogated for a geological surface, thus allowing the extraction of the 3D object to be automated and eliminating the need to manually digitize surfaces. Since the function is continuous throughout space and does not depend on a mesh or grid for its definition, the extracted geological wireframe can be constructed at any desired resolution in the specific volume of interest [67]. Nevertheless, it must be considered that direct implicit modeling yielded better fitting near the constraint line but worse fitting far from the constraint line [68], and where traditional wireframing allows the user to manipulate the modeling process, on a local scale to overcome data density issues, the implicit modeling process is entirely reliant on the input data accuracy and the modeling parameters for the geological interpretation.

4. Results

4.1. Electrical Resistivity Models

Two inversion methods have been used for the P1 to P7 ERT cross-sections: smoothing-constrained and robust. In all cases, the results of the mathematical inversion process have been satisfactory, as the convergence criterion used (root mean square or RMS) has values always below 5%. However, after comparing the two sections obtained in each case, the smooth method was chosen for interpreting ERT results and for modeling. The smoothing-constrained method has shown a more consistent geological interpretation and better definition of the geometry of the identified lithological units, especially the contact between the aquifer and the basement.

The electrical resistivity values obtained from the inversion of the 17 ERT cross-sections mainly ranged between 13 and 5000 Ω·m. From geoelectrical records, three layers can be distinguished according to their electrical resistivity values. The shallowest layer is characterized by resistivity values higher than 600 Ω·m and can be identified at the upper part of geoelectrical cross-sections. The level is interpreted as gravels and sands from the shallow aquifer above the water table and has a thickness always identified below 15 m. At this layer, there were important lateral variations in the resistivity values. These variations reflect lateral changes in the facies, which are usually transitions to fine-grained sediments.

Below, the ERT cross-sections show a unit of variable thickness from 3 to 45 m thickness characterized by 20–600 Ω·m resistivity values. These electrical resistivity values are interpreted as fine particles or saturated sediments from the shallow aquifer.

The third unit is characterized by high resistivity values in the 600–2500 Ω·m range and corresponds to the schists of the basement and their top limit has been used to infer the aquifer/aquiclude contact. The contact is identified as a long amplitude irregular surface in most of the cases, but in particular, some sections show a stepped morphology as can be identified in Figure 5.

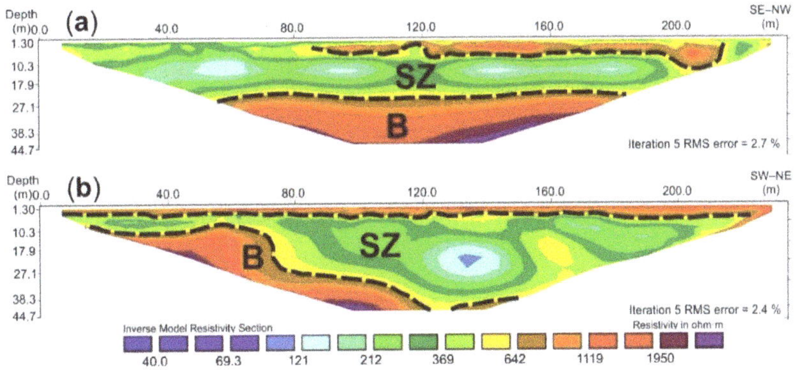

Figure 5. Examples of electrical resistivity tomography (ERT) cross-section results obtained at Port de la Selva site. (**a**) ERT cross-section P8; (**b**) ERT cross-section P1. B: bedrock (aquiclude unit); SZ: Saturated shallow aquifer; black dashed line represents the inferred hydrogeological contact.

4.2. Aquifer Geometry

The 6342 electrical resistivity values obtained from the ERT cross-sections and data of lithological logs from six boreholes were used to delineate the geometry of the contact between the shallow aquifer and the basement and to infer the potential aquifer thickness all over the studied area. Lithological logs from water-supply wells or other boreholes gathered were not used as they do not reach the contact depth. The area characterized was 10 km^2 and the volume of the aquifer estimated by the model is close to $180 \cdot 10^6$ m^3.

The geometry of the contact between the aquifer and the metamorphic bedrock is quite irregular and is located at an average depth of about 18 m. The maximum depth contact or equivalent aquifer's maximum thickness (53 m) is located at the mouth of Romanyanc Stream and close to the municipal water supply wells. It shows a progressive increase of thickness along the Selva Stream path towards the Port de la Selva beach. The general trend is also identified in the Riera de Romanyac, up to about 400 m downstream from the position of SAT facilities. From this point, the base morphology becomes more irregular and values close to 35 m thickness are obtained within the SAT ponds facilities and more densely ERT and borehole data area (Figure 6). The volume modeling method has worked then on scattered drill hole data of any data density, including processing combined information from dense control data in SAT facilities area (providing higher resolution) as well as sparse resource drilling outside SAT data (showing more smoothed boundaries).

4.3. Saltwater Intrusion

According to the authors of [69], the electrical resistivity values of the saturated zone in alluvial coastal aquifers range from 10 to 100 Ohm·m, depending on the total dissolved solids (TDS) concentration. The TDS content in groundwater is an indication of its salinity and the electrical resistivity decreases progressively with an increase in the levels of ionic concentrations or salinity in groundwater. The presence of a low resistivity zone in ERT sections (lower than 10 Ohm·m) can be interpreted as indicating the presence of a seawater intrusion [70–72] because freshwater typically has a resistivity of between 50 and 100 Ohm·m [73], with a resistivity of 10 to 50 Ohm·m corresponding to the transition, or brackish water zone [32]. In terms of the study area, we have selected values smaller than 10 Ohm·m for mapping the saltwater intrusion and in the range of 10–50 Ohm·m to infer a transition zone using implicit modeling tools.

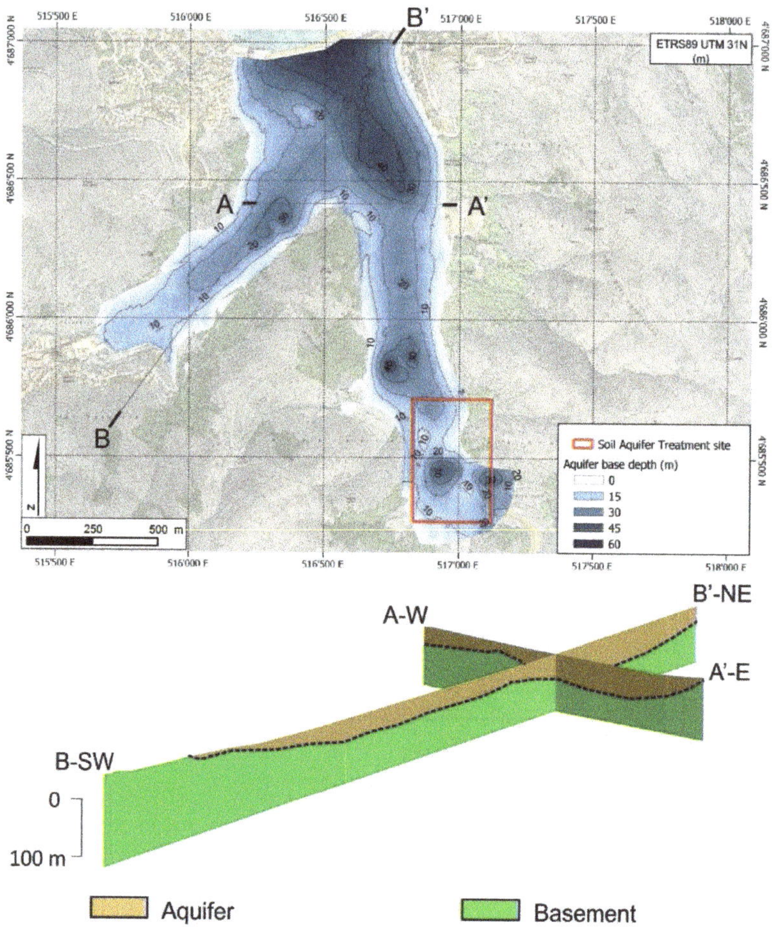

Figure 6. Shallow aquifer base depth model inferred from ERT and borehole data (Port de la Selva, Spain).

The 3D implicit model obtained shows a general resistivity increasing trend towards the hills. An inward feature wedge-shaped with low resistivity values (below 10 Ohm·m) is identified at the seaside—until 310 m inland—which highlights possible marine intrusion (Figure 7). Resistivities in the range of 10–100 Ohm·m are also mainly located close to the coastline, reach 400 m inland, and should include a brackish or mixed water zone. The model also showed a low resistivity anomaly 1–100 Ω·m beneath and towards the location of the municipality water-supply well reflecting a low resistivity anomaly as a result of a rising deep saline water effect (upcoming) from intensive water pumping.

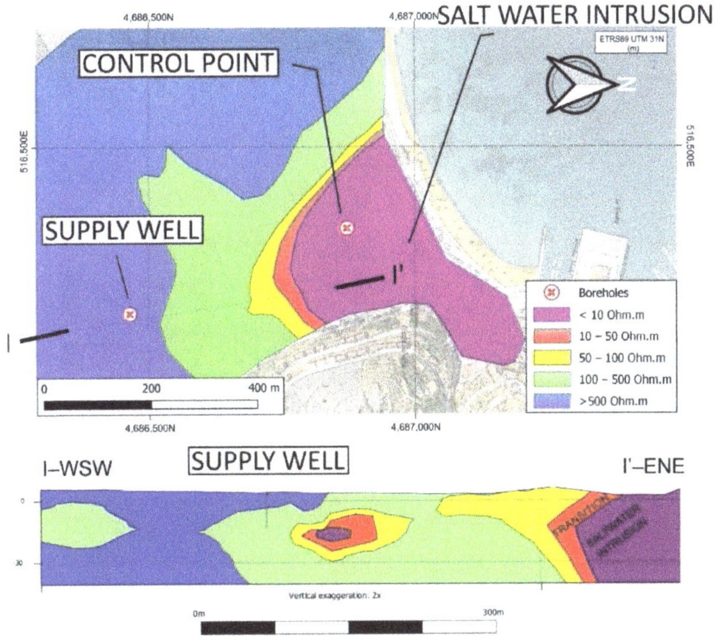

Figure 7. 3D resistivity plain view and cross-section model of the study area inferred from ERT data. Resistivity values lower than 10 Ohm.m are assumed as a threshold of seawater intrusion delineation. Distances are in meters.

5. Discussion

5.1. Aquifer Geometry

Aquifer geometry assessment was performed using two conceptual hydrogeological cross-sections from the ICGC [42,43] and comparing data from the detailed 2D ERT cross-sections and lithological logs of boreholes from SAT facilities project.

Among the geometry model (Figure 6) and hydrogeological cross-sections (Figure 3), there are thickness and general morphology discrepancies. However, the model shows a similar contact morphology when compared to longitudinal geological cross-section and similar maximum depth when comparing to transverse conceptual section. In both directions, average aquifer thickness is consistent.

Based on the information from the lithological logs of research piezometers, we identify also issues in correlating geoelectrical response and lithological logs if boreholes are more than eight meters apart. As an example, the ERT cross-section P12 results are easy to correlate with the lithological log of piezometer S3 located eight meters apart. We could identify an increment in resistivity values at two meters above sea level (masl) and a lithological change at the same position. On the other hand, the P11 ERT cross-section and the borehole S6 have 40 m distance among them. The ERT results show a resistivity change at five masl and borehole log S6 identifies a lithological change at three meters below sea level (Figure 8). This fact is probably due to high heterogeneity and non-regular contact between aquifer/aquiclude in the studied area and the majority of classical quantitative hydrogeophysical studies do not specify the issue as they have been performed at the local scale (~10 m), where the scale disparity between direct (wellbore) and indirect (geophysical) measurements is often not significant [74].

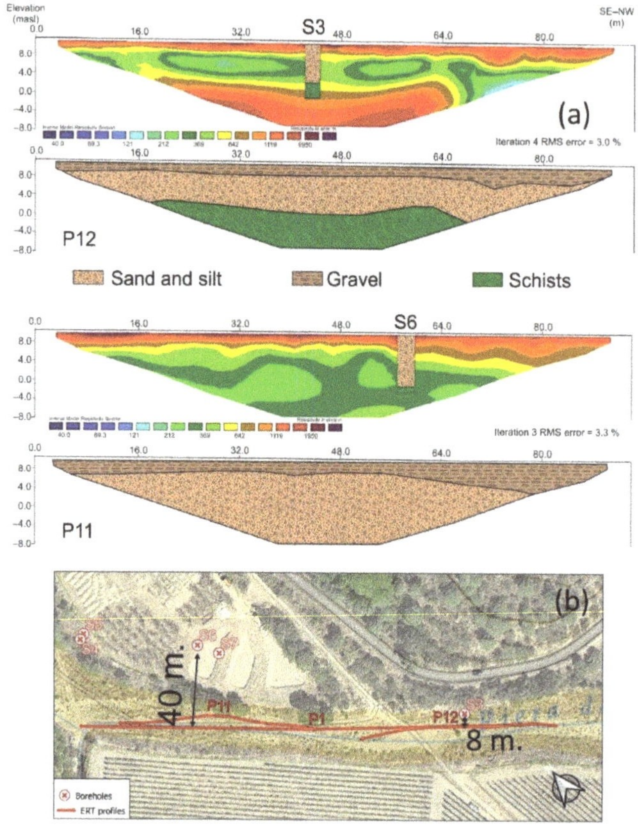

Figure 8. (**a**) P11 and P12 inverted resistivity cross-section and its lithological interpretation; (**b**) location sketch of boreholes S3 and S8 and ERT cross-sections P11 and P12 (Port de la Selva).

As stated by de Marsilly [75], subsurface imaging is a convenient asset, as it can help describe the geometry of a heterogeneous geological system and as previously stated by the authors of [76,77], small-scale data obtained using only wellbore-based methods and in any case simple correlation of layers identified in stratigraphic logs may not provide information about the full 3D geometries of the aquifers to be reconstructed.

5.2. Saltwater Intrusion

Previous modeling studies on the area were focused on simulating aquifer response to rainfall events and pumping in water-supply wells and testing the sensitivity of travel time and dilution rate to several aspects such as rainfall scenarios, infiltration rates, pumping rates in water supply wells, and (uncertain) aquifer parameters such as porosity and hydraulic conductivity [35]. Specific studies delineating saltwater intrusion have not been previously published. The delineation generally requires multiple depth sampling at different water control points or implies combining flow and solute transport equations which are not easy to model even though numerically.

The saltwater intrusion model was obtained by combining ERT data from two different acquisition campaigns and unveiled an image of the issue. The model of Port de la Selva's shallow aquifer shows an area of low resistivity values close to the coastline. The existence of current groundwater salinity at this area is confirmed by the results provided by the town council from supply water wells and the Catalan Water Agency (ACA) from the groundwater control point (Figure 7). The control point is sampled once a year—in September—coinciding with the end of the main holiday season. The indicators of salinity used are some ionic concentrations and the electrical conductivity of water samples (Table 2). Additionally, chloride concertation's up to 700 mg/L have been reported during summer 2018. Therefore, it was necessary to contract a portable desalination plant to guarantee standard quality for drinking water. Periodically, groundwater analysis clearly exceed the local regulations for human water supply without any additional treatment and the range for being considered freshwater according to World Health Organization [31].

Table 2. Salinity indicators results of Port de la Selva aquifer control point compared to World Health Organization suggested groundwater classification and Spanish sanitary guidelines for the quality of water for human consumption.

Physicochemical Parameter	Units	2017	2018	2019	WHO Range for Aquifers [1]			RD 14/2013 [2]
					Fresh	Brackish	Saline	
Na^+	(mg/L)	361	>1000	>1000	<200	200–400	>400	200
K^+	(mg/L)	9	67	30	<55	55–70	>70	—
Cl^-	(mg/L)	683	8518	4329	<250	250–1000	>1000	250
SO_4^{2-}	(mg/L)	129	1008	536	<200	200–500	>500	250
EC [3]	(μS/cm)	2315	24,271	12,342	<1500	1500–3000	>3000	2500

[1] World Health Organization (WHO) suggested groundwater classification [78]. [2] Spanish sanitary guidelines for the quality of water for human consumption [37]. [3] Laboratory electrical conductivity at 20 °C.

However, the delineation must be used as a qualitative image of the saltwater intrusion and to assist the straightforward design of optimal and cost-effective acquisition future field surveys. At this stage, the use of a constant resistivity threshold to improve the delineation of seawater intrusion must include additional information about the geological heterogeneity of the aquifer and direct estimation of TDS in the low salinity (high resistivity) region should be avoided as it is highly sensitive to clay content which is not properly defined [22].

Future research results could be fast incorporated such as other data from other geophysical techniques such as electromagnetic techniques, and seismic methods successfully applied as by the authors of [21] in characterizing coastal shallow aquifers and new research boreholes specifically drill on Port de la Selva coastal area for saltwater delineation scope. Moreover, the use of the ERT methodology could be implemented to other shallow coastal aquifer sites for qualitative saltwater identification and monitoring and in groundwater models used for decision-making management of freshwater resources.

Finally, we are convinced that the methodology has yielded information that can greatly help us to define better the aquifer salinization extension and apply the geometry coupled with numerical groundwater model, major ions, and isotope data to determine travel time and dilution rate with more acceptable reliability [38]. Furthermore, in projects where it is necessary to follow up on a timeline, it would be possible to re-evaluate new conditions (salinity, phreatic changes, new geophysics data, water environment changes . . .) and remediation or response time could be drastically reduced. All the proposed approaches could improve the commitment of stakeholders to the benefits of SAT and/or adopting the methodology on a larger scale.

6. Conclusions

The geometry of the Port de la Selva shallow aquifer is irregular and the correlation between the data from boreholes and electrical resistivity tomography shows that the latter geophysical prospecting technique is a suitable tool for in-depth analysis of shallow aquifers geometry. It provides a high-resolution geological correlation as well as uninterrupted monitoring of the thickness of aquifer zones. The 17 obtained geoelectric cross-sections had variable values of resistivity, both laterally and in-depth. In the vertical dimension, three layers can be distinguished. The superficial level had high resistivity values, which correlated with gravels and sands above the water table. The second level had very low resistivity values, corresponding to water-saturated aquifer sediments. The thickness of this intermediate area is highly variable, ranging from 3 to 45 m at nearby points. Finally, the lowest level corresponds to the Paleozoic substrate and was characterized by a progressive increase in resistivity values.

The lithological data from borehole and resistivity data from ERT have been progressively and fast incorporated into the 3D aquifer geometry and saltwater intrusion models using implicit modeling tools. We have used implicit modeling software for deriving the aquifer geometry model from discrete variables such as lithological information and ground resistivity and a saltwater intrusion model from the same resistivity data.

The aquifer geometry model shows the contact between the shallow aquifer and aquiclude unit and infer the potential aquifer thickness all over the studied area. The geometry of the contact between the aquifer and the metamorphic bedrock is quite irregular and located at an average depth of about 18 m. The 3D saltwater intrusion model shows a general electrical resistivity increasing trend towards the hills. An inward feature wedge-shaped with low resistivity values (below 10 Ohm·m) is identified at the seaside which highlights possible marine intrusion. Electrical resistivities in the range of 10–100 Ohm·m are also mainly located close to the coastline and should include a brackish or mixed water zone. The model also showed a low resistivity anomaly 1–100 Ω·m beneath and towards the municipality well position reflecting an upcoming that is believed to be triggered by water over-pumping.

The geological model obtained can easily incorporate data from other complementary geophysical techniques, such as electromagnetic and seismic methods. However, we believe that the methodology applied has provided fundamental information on aquifer geometry and the extent of saltwater intrusion, which can greatly help improve the reliability of mathematical models of the aquifer and more accurately determine travel time and dilution rate for artificial recharge projects.

Author Contributions: Conceptualization, A.C. and M.H.; methodology, A.U., A.S., L.R., C.A., and R.G.-A.; formal analysis, A.U., M.H., A.C., and A.S.; data curation, A.U., C.A., R.L., M.H., and A.S.; writing—original draft preparation, A.S. and A.C.; writing—review and editing, A.C., M.H., A.U., C.A., R.L., J.C.T., L.R., R.G.-A., and A.S. All authors have read and agreed to the published version of the manuscript.

Funding: This research received no external funding.

Acknowledgments: This work was supported by Seequent providing a License for Leapfrog Geo™ for the research. Thanks are also due to Lluís Sala (Consorci de la Costa Brava), Xavier Martínez (Centre Tecnològic de Manresa), Andreu Masferrer (SOREA), Pere Ribot (Port de la Selva Municipality) and Martí Bayer (Amphos 21) for their continuous assistance and for providing invaluable information about hydrogeological and water management data in the study area. We thank Albert Pérez and Sebastián González for their considerable contribution to the collection of field data.

Conflicts of Interest: The authors declare no conflict of interest.

References

1. Van der Gun, J. *Groundwater and Global Change: Trends, Opportunities and Challenges | International Groundwater Resources Assessment Centre*; United Nations Educational, Scientific and Cultural Organization: Paris, France, 2012; ISBN 9789230010492.
2. Aeschbach-Hertig, W.; Gleeson, T. Regional strategies for the accelerating global problem of groundwater depletion. *Nat. Geosci.* **2012**, *5*, 853–861. [CrossRef]
3. Akbarpour, S.; Niksokhan, M.H. Investigating effects of climate change, urbanization, and sea level changes on groundwater resources in a coastal aquifer: An integrated assessment. *Environ. Monit. Assess.* **2018**, *190*. [CrossRef]
4. Mastrocicco, M.; Colombani, N. The Issue of Groundwater Salinization in Coastal Areas of the Mediterranean Region: A Review. *Water* **2021**, *13*, 90. [CrossRef]
5. Custodio, E. Coastal aquifers of Europe: An overview. *Hydrogeol. J.* **2010**, *18*, 269–280. [CrossRef]
6. Alfarrah, N.; Walraevens, K. Groundwater overexploitation and seawater intrusion in coastal areas of arid and semi-arid regions. *Water* **2018**, *10*, 143. [CrossRef]
7. Sabater, S.; Barceló, D. *Water Scarcity in the Mediterranean: Perspectives under Global Change*; Springer: Berlin, Germany, 2010, ISBN 978-3-642-03970-6(H).

8. Marecos, M.H. *Water reuse in Europe*; Official Publication of the European Water Association (EWA): Hennef, Germany, 2007.
9. Dillon, P.; Escalante, E.F.; Megdal, S.B.; Massmann, G. Managed aquifer recharge for water resilience. *Water* **2020**, *12*, 1846. [CrossRef]
10. Angelakis, A.N.; Marecos Do Monte, M.H.F.; Bontoux, L.; Asano, T. The status of wastewater reuse practice in the Mediterranean basin: Need for guidelines. *Water Res.* **1999**, *33*, 2201–2217. [CrossRef]
11. Dillon, P.; Pavelic, P.; Toze, S.; Rinck-Pfeiffer, S.; Martin, R.; Knapton, A.; Pidsley, D. Role of aquifer storage in water reuse. *Desalination* **2006**, *188*, 123–134. [CrossRef]
12. Elkayam, R.; Sopliniak, A.; Gasser, G.; Pankratov, I.; Lev, O. Oxidizer Demand in the Unsaturated Zone of a Surface-Spreading Soil Aquifer Treatment System. *Vadose Zone J.* **2015**, *14*, 1–10. [CrossRef]
13. Bekele, E.; Page, D.; Vanderzalm, J.; Kaksonen, A.; Gonzalez, D. Water recycling via aquifers for sustainable urban water quality management: Current status, challenges and opportunities. *Water* **2018**, *10*, 457. [CrossRef]
14. Burris, D. *Groundwater Replenishment System, 2019 Annual Report*; Orange County Water District: Fountain Valley, CA, USA, 2020.
15. De Carlo, L.; Caputo, M.C.; Masciale, R.; Vurro, M.; Portoghese, I. Monitoring the drainage efficiency of infiltration trenches in fractured and karstified limestone via time-lapse hydrogeophysical approach. *Water* **2020**, *12*, 2009. [CrossRef]
16. Bekele, E.; Toze, S.; Patterson, B.; Higginson, S. Managed aquifer recharge of treated wastewater: Water quality changes resulting from infiltration through the vadose zone. *Water Res.* **2011**, *45*, 5764–5772. [CrossRef] [PubMed]
17. Council, N.R. *Ground Water Recharge Using Waters of Impaired Quality*; The National Academies Press: Washington, DC, USA, 1994; ISBN 978-0-309-05142-2.
18. Gale, I. *Strategies for Managed Aquifer Recharge (MAR) in Semi-Arid Areas*; International Association of Hydrogeologists commission on Management of Aquifer Recharge IAH-MAR, UNESCO Division of Water Sciences: Paris, France, 2005.
19. Ringleb, J.; Sallwey, J.; Stefan, C. Assessment of managed aquifer recharge through modeling—A review. *Water* **2016**, *8*, 579. [CrossRef]
20. EWRI. *Standard Guidelines for Artificial Recharge of Ground Water*; Environmental and Water Resources Institute, Ed.; Standards; American Society of Civil Engineers (ASCE): Reston, VA, USA, 2001; ISBN 978-0-7844-0548-2.
21. Paz, M.C.; Alcalá, F.J.; Medeiros, A.; Martínez-Pagán, P.; Pérez-Cuevas, J.; Ribeiro, L. Integrated MASW and ERT imaging for geological definition of an unconfined alluvial aquifer sustaining a coastal groundwater-dependent ecosystem in southwest Portugal. *Appl. Sci.* **2020**, *10*, 5905. [CrossRef]
22. González-Quirós, A.; Comte, J.C. Relative importance of conceptual and computational errors when delineating saltwater intrusion from resistivity inverse models in heterogeneous coastal aquifers. *Adv. Water Resour.* **2020**, *144*, 103695. [CrossRef]
23. Behroozmand, A.A.; Auken, E.; Knight, R. Assessment of Managed Aquifer Recharge Sites Using a New Geophysical Imaging Method. *Vadose Zone J.* **2019**, *18*, 1–13. [CrossRef]
24. Binley, A.; Hubbard, S.S.; Huisman, J.A.; Revil, A.; Robinson, D.A.; Singha, K.; Slater, L.D. The emergence of hydrogeophysics for improved understanding of subsurface processes over multiple scales. *Water Resour. Res.* **2015**, *51*, 3837–3866. [CrossRef]
25. Drahor, M.G.; Berge, M.A.; Göktürkler, G.; Kurtulmuş, T.Ö. Mapping aquifer geometry using electrical resistivity tomography: A case study from Şanlıurfa, south-eastern Turkey. *Near Surf. Geophys.* **2011**, *9*, 55–66. [CrossRef]
26. Benabdelouahab, S.; Salhi, A.; Himi, M.; Stitou El Messari, J.E.; Casas Ponsati, A. Geoelectrical investigations for aquifer characterization and geoenvironmental assessment in northern Morocco. *Environ. Earth Sci.* **2019**, *78*, 1–16. [CrossRef]
27. Sendrós, A.; Himi, M.; Lovera, R.; Rivero, L.; Garcia-Artigas, R.; Urruela, A.; Casas, A. Geophysical Characterization of Hydraulic Properties around a Managed Aquifer Recharge System over the Llobregat River Alluvial Aquifer (Barcelona Metropolitan Area). *Water* **2020**, *12*, 3455. [CrossRef]
28. Martínez, J.; Benavente, J.; García-Aróstegui, J.L.; Hidalgo, M.C.; Rey, J. Contribution of electrical resistivity tomography to the study of detrital aquifers affected by seawater intrusion-extrusion effects: The river Vélez delta (Vélez-Málaga, southern Spain). *Eng. Geol.* **2009**, *108*, 161–168. [CrossRef]
29. Galazoulas, E.C.; Mertzanides, Y.C.; Petalas, C.P.; Kargiotis, E.K. Large Scale Electrical Resistivity Tomography Survey Correlated to Hydrogeological Data for Mapping Groundwater Salinization: A Case Study from a Multilayered Coastal Aquifer in Rhodope, Northeastern Greece. *Environ. Process.* **2015**, *2*, 19–35. [CrossRef]
30. Kazakis, N.; Pavlou, A.; Vargemezis, G.; Voudouris, K.S.; Soulios, G.; Pliakas, F.; Tsokas, G. Seawater intrusion mapping using electrical resistivity tomography and hydrochemical data. An application in the coastal area of eastern Thermaikos Gulf, Greece. *Sci. Total Environ.* **2016**, *543*, 373–387. [CrossRef]
31. Hasan, M.; Shang, Y.; Jin, W.; Shao, P.; Yi, X.; Akhter, G. Geophysical Assessment of Seawater Intrusion into Coastal Aquifers of Bela Plain, Pakistan. *Water* **2020**, *12*, 3408. [CrossRef]
32. Vann, S.; Puttiwongrak, A.; Suteerasak, T.; Koedsin, W. Delineation of seawater intrusion using geo-electrical survey in a coastal aquifer of Kamala beach, Phuket, Thailand. *Water* **2020**, *12*, 506. [CrossRef]
33. Lajaunie, C.; Courrioux, G.; Manuel, L. Foliation fields and 3D cartography in geology: Principles of a method based on potential interpolation. *Math. Geol.* **1997**, *29*, 571–584. [CrossRef]
34. SMC. *Climatologia. L'Alt Empordà. 1961–2000*; Servei Meteorològic de Catalunya—Departament de Medi Ambient i Habitatge, Generalitat de Catalunya: Barcelona, Spain, 2014.
35. Bayer-Raich, M.; Vilanova, E.; Jordana, S. Water reuse in El Port de la Selva: Groundwater modelling of Soil Aquifer Treatment (SAT) for reclaimed water. In Proceedings of the 5th International Symposium RE-WATER Braunschweig, Braunschweig, Germany, 2–3 November 2015; pp. 158–174.

36. Fajnorová, S.; Sprenger, C.; Hermes, N.; Ternes, T.A.; Sala, L.; Miehe, U.; Drewes, J.E.; Hübner, U. Assessment of Full-Scale Indirect Potable Water Reuse in El Port de la Selva, Spain. *Water* **2021**, *13*, 325. [CrossRef]
37. MPR Real Decreto 140/2003, de 7 de Febrero, por el que se Establecen los Criterios Sanitarios de la Calidad del Agua de Consumo Humano. Ministerio de la Presidencia Gobierno de España. Available online: http://www.boe.es/boe/dias/2003/02/21/pdfs/A07228-07245.pdf (accessed on 21 January 2021).
38. Martinez, X.; Hochstrat, R.; Miehe, U.; Casado, F.; Frijns, J.; Bortoli, J.; Denieul, M.P.; Orsoni, J.; Alcalde, L.; Jeffrey, J.; et al. *Final Report: DEMOWARE (Innovation Demonstration for a Competitive and Innovative European Water Reuse Sector)*; Community Research and Development Information Service (CORDIS): Luxembourg, 2017.
39. Sprenger, C.; Kraus, F.; Schwarzmüller, H.; Miehe, U.; Bayer, M.; Vilanova, E. Pretreatment requirements and design guidelines for SAT technologies, and two SAT case studies. El Port de la Selva. In *DEMOWARE (Innovation Demonstration for a Competitive and Innovative European Water Reuse Sector)*; Community Research and Development Information Service (CORDIS): Luxembourg, 2017; pp. 100–163.
40. Druguet, E. The Structure of the NE Cap de Creus Peninsula Relationships with Metamorphism and Magmatism. Ph.D Thesis, Universitat Autònoma de Barcelona, Barcelona, Spain, 1997.
41. ICGC. *Alt Empordà. Mapa Geològic Comarcal de Catalunya 1:50 000*; Institut Cartogràfic i Geològic de Catalunya: Barcelona, Spain, 2006.
42. ICGC. *Mapa Hidrogeològic de Roses, Cap de Creus i Far de Creus 1:25 000*; Institut Cartogràfic i Geològic de Catalunya: Barcelona, Spain, 2013.
43. ICGC. *Mapa Hidrogeològic de Llança 1:25 000*; Institut Cartogràfic i Geològic de Catalunya: Barcelona, Spain, 2013.
44. Dahlin, T. 2D resistivity surveying for environmental and engineering applications. *First Break* **1996**, *14*, 275–284. [CrossRef]
45. Sasaki, Y. Resolution of Resistivity Tomography inferred from numerical simulation. *Geophys. Prospect.* **1992**, *40*, 453–463. [CrossRef]
46. Loke, M.H.; Barker, R.D. Rapid least-squares inversion of apparent resistivity pseudosections by a quasi-Newton method. *Geophys. Prospect.* **1996**, 131–152. [CrossRef]
47. Loke, M.H.; Dahlin, T. A comparison of the Gauss–Newton and quasi-Newton methods in resistivity imaging inversion. *J. Appl. Geophys.* **2002**, *49*, 149–162. [CrossRef]
48. Ellis, R.G.; Oldenburg, D.W. Applied geophysical inversion. *Geophys. J. Int.* **1994**, *116*, 5–11. [CrossRef]
49. Oyeyemi, K.D.; Aizebeokhai, A.P.; Adagunodo, T.A.; Olofinnade, O.M.; Sanuade, O.A.; Olaojo, A.A. Subsoil characterization using geoelectrical and geotechnical investigations: Implications for foundation studies. *Int. J. Civ. Eng. Technol.* **2017**, *8*, 302–314.
50. Everett, M.E. Electrical resistivity method. In *Near-Surface Applied Geophysics*; Cambridge University Press: Cambridge, UK, 2013; pp. 70–103. ISBN 9781139088435.
51. Martorana, R.; Fiandaca, G.; Casas Ponsati, A.; Cosentino, P.L. Comparative tests on different multi-electrode arrays using models in near-surface geophysics. *J. Geophys. Eng.* **2009**, *6*, 101–110. [CrossRef]
52. Dahlin, T.; Zhou, B. A numerical comparison of 2D resistivity imaging with 10 electrode arrays. *Geophys. Prospect.* **2004**, *52*, 379–398. [CrossRef]
53. Rey, J.; Martínez, J.; Mendoza, R.; Sandoval, S.; Tarasov, V.; Kaminsky, A.; Hidalgo, M.C.; Morales, K. Geophysical characterization of aquifers in southeast spain using ERT, TDEM, and vertical seismic reflection. *Appl. Sci.* **2020**, *10*, 7365. [CrossRef]
54. Seequent. *User Manual for Leapfrog Geo version 6.0*; Seequent Limited: Christchurch, New Zealand, 2020.
55. Hardy, R.L. Multiquadric equations of topography and other irregular surfaces. *J. Geophys. Res.* **1971**, *76*, 1905–1915. [CrossRef]
56. Carlson, R.E.; Foley, T.A. Interpolation of track data with radial basis methods. *Comput. Math. Appl.* **1992**, *24*, 27–34. [CrossRef]
57. Franke, R. Scattered Data Interpolation: Tests of Some Method. *Math. Comput.* **1982**, *38*, 181–200. [CrossRef]
58. Piret, C.; Dissanayake, N.; Gierke, J.S.; Fornberg, B. The Radial Basis Functions Method for Improved Numerical Approximations of Geological Processes in Heterogeneous Systems. *Math. Geosci.* **2020**, *52*, 477–497. [CrossRef]
59. Cowan, E.J.; Beatson, R.K.; Fright, W.R.; Mclennan, T.J.; Mitchell, T.J. Rapid geological modelling. In Proceedings of the Applied Structural Geology for Mineral Exploration and Mining, International Symposium, Kalgoorlie, Australia, 23–25 September 2002; pp. 1–9.
60. Alhuri, Y.; Ouazar, D.; Taik, A. Radial Basis Functions Alternative Solutions to Shallow Water Equations. *Front. Sci. Eng.* **2011**, *1*, 25–40.
61. Scheuerer, M.; Schaback, R.; Schlather, M. Interpolation of spatial data-A stochastic or a deterministic problem? *Eur. J. Appl. Math.* **2013**, *24*, 601–629. [CrossRef]
62. Hillier, M.J.; Schetselaar, E.M.; de Kemp, E.A.; Perron, G. Three-Dimensional Modelling of Geological Surfaces Using Generalized Interpolation with Radial Basis Functions. *Math. Geosci.* **2014**, *46*, 931–953. [CrossRef]
63. Houlding, S.W. *3D Geoscience Modeling*; Springer: Berlin/Heidelberg, Germany, 1994; ISBN 978-3-642-79014-0.
64. Vollgger, S.A.; Cruden, A.R.; Ailleres, L.; Cowan, E.J. Regional dome evolution and its control on ore-grade distribution: Insights from 3D implicit modelling of the Navachab gold deposit, Namibia. *Ore Geol. Rev.* **2015**, *69*, 268–284. [CrossRef]
65. Cowan, E.J. Deposit-scale structural architecture of the Sigma-Lamaque gold deposit, Canada—insights from a newly proposed 3D method for assessing structural controls from drill hole data. *Miner. Depos.* **2020**, *55*, 217–240. [CrossRef]
66. Beatson, R.K.; Cherrie, J.B.; Mouat, C.T. Fast fitting of radial basis functions: Methods based on preconditioned GMRES iteration. *Adv. Comput. Math.* **1999**, *11*, 253–270. [CrossRef]

67. Cowan, E.J.; Beatson, H.J.; Ross, H.J.; Fright, W.R.; McLennan, T.J.; Evans, T.R.; Carr, J.C.; Lane, R.G.; Bright, D.V.; Gillman, A.J.; et al. Practical Implicit Geological Modelling. In Proceedings of the 5th International Mining Geology Conference, Victoria, Australia, 17–19 November 2003; The Australasian Institute of Mining and Metallurgy: Carlton Victoria, Australia, 2003; pp. 89–99.
68. Guo, J.; Wang, J.; Wu, L.; Liu, C.; Li, C.; Li, F.; Lin, M.; Jessell, M.W.; Li, P.; Dai, X.; et al. Explicit-implicit-integrated 3-D geological modelling approach: A case study of the Xianyan Demolition Volcano (Fujian, China). *Tectonophysics* **2020**, *795*, 228648. [CrossRef]
69. Sherif, M.; El Mahmoudi, A.; Garamoon, H.; Kacimov, A.; Akram, S.; Ebraheem, A.; Shetty, A. Geoelectrical and hydrogeochemical studies for delineating seawater intrusion in the outlet of Wadi Ham, UAE. *Environ. Geol.* **2006**, *49*, 536–551. [CrossRef]
70. Abdul Nassir, S.S.; Loke, M.H.; Lee, C.Y.; Nawawi, M.N.M. Salt-water intrusion mapping by geoelectrical imaging surveys. *Geophys. Prospect.* **2000**, *48*, 647–661. [CrossRef]
71. Song, S.H.; Lee, J.Y.; Park, N. Use of vertical electrical soundings to delineate seawater intrusion in a coastal area of Byunsan, Korea. *Environ. Geol.* **2007**, *52*, 1207–1219. [CrossRef]
72. Chafouq, D.; El Mandour, A.; Elgettafi, M.; Himi, M.; Bengamra, S.; Lagfid, Y.; Casas, A. Assessing of Saltwater Intrusion in Ghiz-Nekor Aquifer (North Morocco) Using Electrical Resistivity Tomography. In Proceedings of the Near Surface Geoscience 2016—22nd European Meeting of Environmental and Engineering Geophysics, Barcelona, Spain, 4–8 September 2016; Volume 2016. Available online: https://doi.org/10.3997/2214-4609.201602066 (accessed on 30 December 2020).
73. Kura, N.U.; Ramli, M.F.; Ibrahim, S.; Sulaiman, W.N.A.; Zaudi, M.A.; Aris, A.Z. A preliminary appraisal of the effect of pumping on seawater intrusion and upconing in a small tropical island using 2D resistivity technique. *Sci. World J.* **2014**, *2014*. [CrossRef] [PubMed]
74. Rubin, Y.; Hubbard, S.S. *Hydrogeophysics*; Springer: Dordrecht, The Netherlands, 2005; ISBN 978-1-4020-3101-4.
75. de Marsily, G.; Delay, F.; Gonçalvès, J.; Renard, P.; Teles, V.; Violette, S. Dealing with spatial heterogeneity. *Hydrogeol. J.* **2005**, *13*, 161–183. [CrossRef]
76. Chen, J.; Hubbard, S.S.; Gaines, D.; Korneev, V.; Baker, G.; Watson, D. Stochastic estimation of aquifer geometry using seismic refraction data with borehole depth constraints. *Water Resour. Res.* **2010**, *46*, W11539. [CrossRef]
77. Ghiglieri, G.; Buttau, C.; Arras, C.; Funedda, A.; Soler, A.; Barbieri, M.; Carrey, R.; Domènech, C.; Torrentó, C.; Otero, N.; et al. Using a multi-disciplinary approach to characterize groundwater systems in arid and semi-arid environments: The case of Biskra and Batna regions (NE Algeria). *Sci. Total Environ.* **2020**, *757*, 143797. [CrossRef] [PubMed]
78. APHA. *Standard Methods for the Examination of Water and Wastewater*, 20th ed.; Clesceri, L., Greenberg, A., Eaton, A., Eds.; American Public Health Association: Washington, DC, USA, 1999, ISBN 0-87553-235-7.

applied sciences

MDPI

Article

Identifying Changes in Sediment Texture along an Ephemeral Gravel-Bed Stream Using Electrical Resistivity Tomography 2D and 3D

Marcos A. Martínez-Segura [1], Carmelo Conesa-García [2], Pedro Pérez-Cutillas [2], Pedro Martínez-Pagán [1] and Marco D. Vásconez-Maza [1,*]

[1] Department of Mining and Civil Engineering, Universidad Politécnica de Cartagena, 52, Paseo Alfonso XIII, 30203 Cartagena, Spain; marcos.martinez@upct.es (M.A.M.-S.); p.martinez@upct.es (P.M.-P.)
[2] Department of Physical Geography, Campus de la Merced, University of Murcia, s/n, 30001 Murcia, Spain; cconesa@um.es (C.C.-G.); pedrope@um.es (P.P.-C.)
* Correspondence: marco.vasconez@edu.upct.es; Tel.: +34-968-327-033

Abstract: Differences in deposit geometry and texture with depth along ephemeral gravel-bed streams strongly reflect fluctuations in bedload which are due to environmental changes at the basin scale and to morphological channel adjustments. This study combines electrical resistivity tomography (ERT) with datasets from borehole logs to analyse the internal geometry of channel cross-sections in a gravel-bed ephemeral stream (southeast Spain). The survey was performed through longitudinal and transverse profiles in the upper channel stretch, of 14 to 30 m in length and 3 to 6 m in depth, approximately. ERT values were correlated with data on sediment texture as grain size distribution, effective grain sizes, sorting, and particle shape (Zingg's classification). The alluvial channel-fills showed the superposition of four layers with uneven thickness and arrangement: (1) the softer rocky substrate (<1000 Ω.m); (2) a thicker intermediate layer (1000 to 2000 Ω.m); and (3) an upper set composed of coarse gravel and supported matrix, ranging above 2000 Ω.m, and a narrow subsurface layer, which is the most resistive (>5000 Ω.m), corresponding to the most recent armoured deposits (gravel and pebbles). The ERT results coupled with borehole data allowed for determining the horizontal and vertical behaviour of the materials in a 3D model, facilitating the layer identification.

Keywords: deposition patterns; climate change; ephemeral gravel-bed stream; electrical resistivity tomography; borehole samples

Citation: Martínez-Segura, M.A.; Conesa-García, C.; Pérez-Cutillas, P.; Martínez-Pagán, P.; Vásconez-Maza, M.D. Identifying Changes in Sediment Texture along an Ephemeral Gravel-Bed Stream Using Electrical Resistivity Tomography 2D and 3D. *Appl. Sci.* **2021**, *11*, 3030. https://doi.org/10.3390/app11073030

Academic Editor: Edoardo Rotigliano

Received: 28 February 2021
Accepted: 22 March 2021
Published: 29 March 2021

1. Introduction

Ephemeral streams are watercourses of arid and semi-arid environments with unstable morphology and high temporal variability of runoff. Sudden, extreme discharge events, that are isolated in time and alternate with long dry periods. These types of streams are particularly sensitive to short-term climatic changes, and human impacts may alter their degree of response, sometimes leading to large morphological adjustments during flash floods [1–3]. As a result, the ephemeral channels show a changing geometry, highly conditioned by differences in slope and textural variations in the bed materials and banks. Often along their upper reaches and on alluvial fans, these channels have a steep slope which promotes a fast hydraulic regime. Under such conditions, and considering the abundant sediment stored within the channel, important transport rates contribute to most of the morphological changes in the channel. This dynamic is especially complex in gravel-bed ephemeral channels, subject to sporadic and torrential transport. A product of this is the mixture of sand, gravel, and pebbles laid in layers of irregular thicknesses and geometries. The bed material also shows high spatial variability in texture between bedforms, between channel reaches, and between the surface and the subsurface. An uncertainty in bedload estimates for this type of streams is largely driven by the inability to characterise arrangement, orientation, and resultant forces of fluvial sediment in river beds. Water working of grains leads to textural differences between areas of the bed through particle sorting, packing, imbrication, mortaring, and the degree of bed armouring.

This study intends to solve the uncertainty related to the spatial variability of the thickness. Therefore, non-destructive electrical resistivity tomography (ERT) imaging in 2D and 3D is used to visualise, quantify, and assess the internal geometry of cross-sections of an ephemeral gravel-bed stream in combination with datasets extracted from borehole samples. The chosen pilot study area is the upper reach of the Rambla de la Azohía, a gravel-bed channel subjected to great recent geomorphological activity.

In general, the electrical resistivity tomography method provides indirect information that need to be validated by direct measurements (borehole samples), this combination offers satisfactory results. Some studies utilised this synergetic combination for identifying specific elements [4,5], determining deposition material paths [6] or even determining the stratigraphy of fluvial channels [7]. Electrical resistivity tomography (ERT) is a non-invasive method of rapid application, which is usually used to generate models of distributions of electrical properties of the subsurface, from which the geological structure and hydrogeological variations can be inferred [7–9]. Additionally, ERT can be employed to map spatial variations of qualities within a mineral deposit in 3D [10,11]. Several authors conducted ERT surveys to study loose materials, assess mine tailings, and so forth [12–15].

ERT can provide spatial mapping models of the subsurface at the site scale, in comparison with intrusive sampling methods, the information from which is only valid for discrete locations. Various studies have focused on the detail achieving high-resolution results with low electrode spacing [16–19]. This method is sensitive to variations in lithological composition with depth and can be used to distinguish between different types of materials, e.g., gravels and pebbles (characterised by high resistivity), or silts and clay (characterised by low resistivity) [20,21].

In the absence of direct porosity data, the objective of this research is to: (i) assess the grain shape, particle size distribution, and sorting in heterogeneous alluvial deposits; (ii) assess the vertical and lateral behaviour by creating a 3D model based on electrical resistivity tomography; and (iii) set a methodology to evaluate this type of ephemeral channels.

2. Study Area: Geomorphological and Climatic Setting

This study was carried out in an upper stretch of the ephemeral gravel-bed channel known as "Rambla de la Azohía". The Rambla de la Azohía is located in the Murcia Region, in southeastern Spain (Figure 1). The climate of the study area has strong seasonal contrasts, providing extreme drought patterns with mean annual precipitation values of 300 mm, and ETP values higher than 140 mm per month in the summer season. The rains are very irregular and intense, which can cause large and rapid floods [10]. The land cover shows a low anthropic impact, associated with extensive scrub areas with shrub canopy typically lower than 20% of total vegetation.

Geomorphological activity is particularly active as a consequence of intense weathering of metamorphic materials (e.g., phyllites, schists, and quartzites) on hillslopes and the installation of a dense network of gullies in the headwater areas that supply large amounts of coarse sediments to the ephemeral channels. High rates of bedload transport and significant morphological adjustments are produced by sporadic torrential flows. Consequently, complex bed-forms and sedimentary structures have developed, also associated with recent changes in the magnitude and frequency of hydrological events.

Often, these bed-forms are composed on poorly sorted sediments with mixed grain sizes (sand, gravel, and pebbles). Deposits of different textures alternate in depth with a general tendency to increase in grain size from the bottom up [11]. This alluvial fill rests mainly on a Miocene marl substrate, the fine texture of which contrasts with that of the overlying coarse detrital materials.

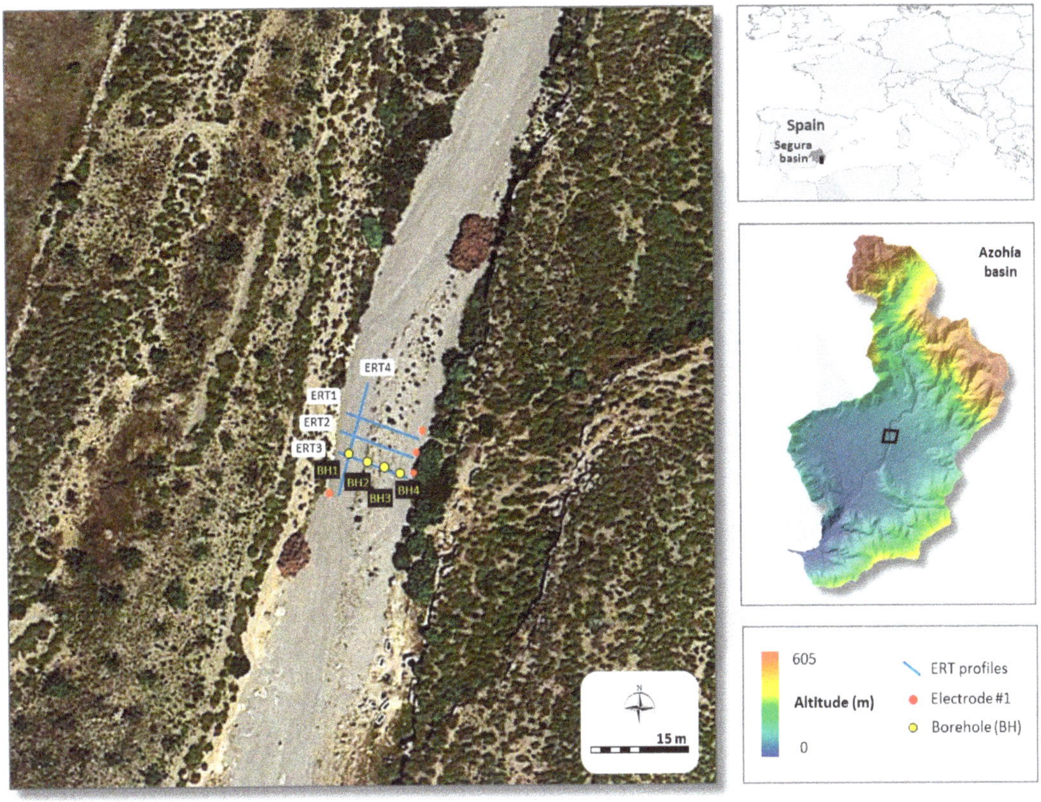

Figure 1. Study area. ERT profiles distribution, red marker determines the beginning of the electrical profile.

3. Materials and Methods

3.1. Electrical Resistivity Tomography (ERT)

Electrical resistivity tomography method is a non-invasive technique. This method is based on a combination of electrical vertical sounding and profiling techniques for gathering subsurface information. ERT uses a basis of four electrodes for acquiring apparent resistivity values in the field. Electrical currents injected into the subsurface employ two electrodes, named current electrodes, which are commonly labelled as A and B. Simultaneously, potential measurements are addressed using another pair of electrodes, named potential electrodes, which are labelled as M and N [12,13]. The four-electrode-based measurement unit principle applies all along the electrical profile, resulting in an over-lapping of layers, allowing for the generation of 2D electrical sections. For 3D models, parallel profiles were conducted, enabling the creation of 3D-format files, which were processed through a next stage by means of an inversion algorithm [14,22].

By using the ERT method, the collection of information from the subsurface will depend on several parameters that are affected by the electrical contrast that defines the subsurface materials. One of these parameters is the investigation depth, where the length of the profile will have a considerable influence on the investigation depth since the more distance is covered by the profile, the deeper the obtained data is [15]. Another key parameter is the type of measurement array, which determines how the measurements will be conducted. Among the different available arrays, this study employed a dipole-dipole array, which offers higher resolution and more detailed imaging. A dipole-dipole array allows for gathering a wider range of measuring points with the same quantity of electrodes [23].

The field data acquisition was conducted with a SuperSting resistivimeter from AGI (Austin, TX, USA) and a gasoline-based power generator from Honda (Valencia, Spain) as a continuous and suitable power supply. The SuperStingTM Wi-Fi R4/IP multielectrode resistivimeter from AGI (Austin, TX, USA) is equipped with 4 channels and 56 stainless steel electrodes, a 56 switch box and measurement resolution for max. 30 nV, which vary according to the voltage level [24].

Figure 1 shows the situation of the ERT profiles and their distributions and covered area. In this case study, a total of four electrical tomography profiles were conducted, where three of which were laid out parallel with a separation of 1.5 m. Those profiles are referred to as ERT1, ERT2, and ERT3, respectively. On the other hand, the ERT profile, named as ERT4, was laid out perpendicular to the others.

ERT4 profile used 56 stainless steel electrodes with a spacing of 0.55 m and a length of 30.25 m. ERT1, ERT2, and ERT3 profiles used 28 stainless steel electrodes with a spacing of 0.55 m and a length of 14.85 m. The investigation depth was ≈6 and 3 m, respectively. Additionally, each electrode was suitably georeferenced using a Leica Zeno 20 GPS equipment (accuracy of ±1 cm) from Leica (Barcelona, Spain).

Before the inversion process, the whole dataset was submitted to pre-processing, consisting of a filtering stage for removing anomalous values and static correction, normalising the variations in resistivity due to the difference in height between electrodes. Then, the terrain topography was incorporated, since any change in the morphology of the terrain affects ERT measurements considerably; and here, the subsequent data inversion was processed [15]. Finally, the dataset was inverted with AGIUSA's EarthImager2D y 3D software (ver. 2.4.4, 1.5.5) which uses the least squares approach [25].

Not only does the software use the finite-element method to solve the inversion, but it also considers the terrain topography. The inversion process computes a model that fits with the observed data. The inversion computes a theoretical initial model which is compared with the measured data. This initial model is progressively modified for reducing discrepancies between the model response and the measured data. The mean absolute misfit error value quantifies the number of differences between models. The iterative process finishes when the computed and measured data achieve an acceptable convergence value. By using L2-norm (smooth), an accurate model of the subsurface of the channel was generated which fitted coherently with the data acquired from the boreholes [22,26].

Soil moisture values, a determining factor for electrical resistivity tests, should be low to minimize artificial variation in resistivity records. To carry out the tomography (prepared on the 3rd of October 2020), the absence of significant rainfall at least one month before the field campaign was considered. In addition, the reference evapotranspiration (ETo) values shown by the nearest meteorological stations provided high values close to 100 mm in September, which, when combined with the high drainage capacity of the sediments of the riverbed, ensured a low humidity of the soil for proper data collection [27].

3.2. *Sediment Texture Analysis from Datasets of Borehole Samples*

Stratigraphic alluvial units were sampled from four boreholes of 3 m depth in the upper channel reach of the Rambla de la Azohía. The drilling sites selected correspond to representative bed forms along this stream stretch: two boreholes were located in a longitudinal alluvial bar and another two in lateral channels (main and secondary). Samples were collected from each sedimentary structure for grain size analysis and the calculation of particle shape (Figure 2).

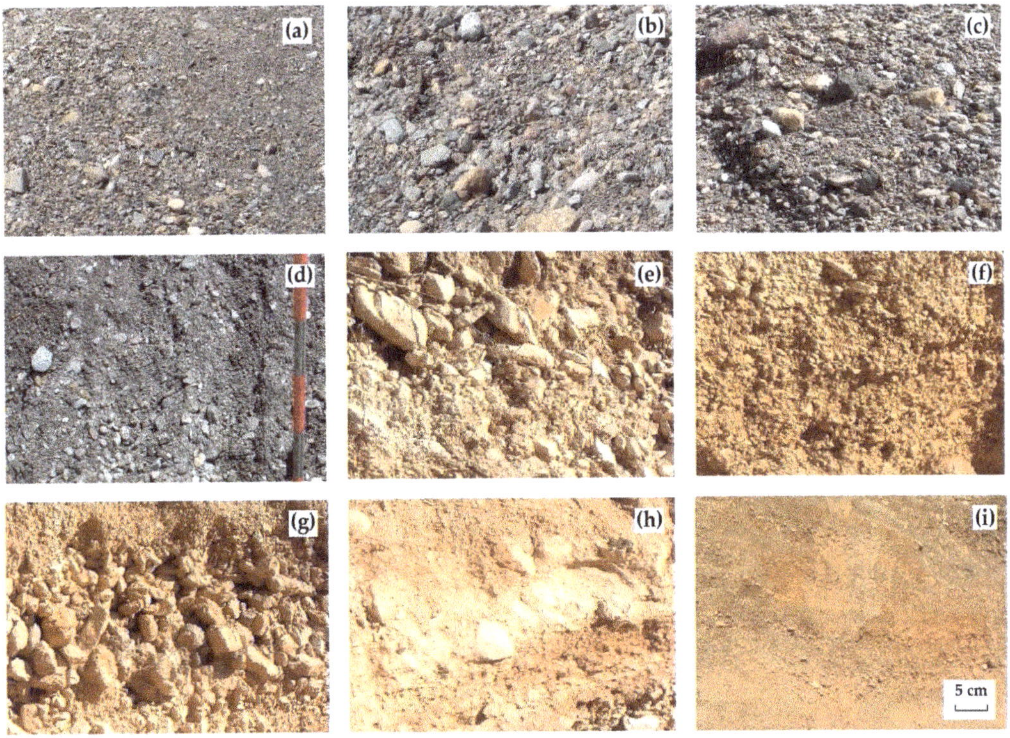

Figure 2. Sediment texture observed on the surface ((**a**): active channel, (**b**): transition area between active channel and central alluvial bar, (**c**): central bar platform) and in the depth of the trench sections ((**d**): sub-surface layer 1 m deep in the active channel area, (**e**–**h**): sub-surface layers 1–1.5 m thick from top to bottom in a bar sedimentary sequence), and (**i**): representative texture of the alluvial fan body cut by the Rambla de la Azohía in its upper reach.

Most of these sediments are unconsolidated materials, whose porosity is closely related to the types of packing, grain shape, arrangement, and particle-size distribution [28]. Finer particles have a more significant effect on porosity than coarse particles [29]. For spherical particles of uniform sizes, the porosity is 47.6% by cubical packing, and 26.0% by rhombohedral packing, which is the most compact packing type [28]. The porosity falls between these two values for other packing classes. However, for porous media in clastic deposits consisting of particles of non-uniform sizes, as in our case, porosity cannot be calculated directly, and other physical properties must be taken into account in addition to the packing effect, such as grain shape, particle size, sorting, and roundness [30]. The primary porosity usually increases as the grain size increases, the sediment is better sorted and more loosely packed, the grains become better rounded, and the clay content decreases [31]. For example, poorly sorted grains generally have closer packing and lower porosity across a wide range of grain sizes, as fine grains tend to fill the void space between large grains. Well-rounded grains with high sphericity should pack with a minimum of pore space [32]. Jin et al. [30] considered four classes of grain packs from different aspect ratios (L/I = 1.0, 1.5, 2.0, 2.5), L/I being defined as the ratio of the length of the longest axis of a grain (L) to that of its shortest axis (I). Its results showed that the spherical grain pack exhibits a higher porosity compared to ellipsoidal grain packs. This effect is also shown in average permeability calculations, where the spherical grain pack exhibits lower permeability. Some typical porosity values of natural sedimentary materials are shown in Table 1.

Table 1. Typical porosity values of natural sedimentary materials [28].

Sedimentary Materials	Porosity (%)	Sedimentary Materials	Porosity (%)
peat soil	60–80	fine-to-medium mixed sand	30–35
soils	50–60	gravel	30–40
clay	45–55	gravel and sand	30–35
silt	40–50	sandstone	10–20
medium-to-coarse mixed sand	35–40	shale	1–10
uniform sand	30–40	limestone	1–10

The median grain size (D_{50}) and 84th percentile (D_{84}) were calculated for each sample. As effective grain sizes were used the 10th (D_{10}) and 20th (D_{20}) percentiles, that is to say, the particle diameter (mm) of 10 and 20% of all sediment is the smallest, respectively by weight. Many authors [33–36] have already regularly used these diameters to calculate hydraulic conductivity (HC) in detritical deposits. The application D_{10} as an effective grain size in this case has been limited to uniform sands [37]. However, in the case of gravels and pebbles, we have found a greater relationship between D_{20} and the ERT results, which can probably be maintained with the HC values. A sorting index (σ) was also adopted as a measure of granulometric dispersion, which takes the expression $\sigma = (D_{84}/D_{16})^{0.5}$.

The derived metrics, particle-shape indices, maximum projection sphericity index (MPSI), and characterisation of sediment textures lead to improved bedload estimates with reduced uncertainty, as well as improved understanding of the relationships between sediment texture, grain size distribution, and electrical resistivity. To calculate the particle-shape, the Zingg (1935) and the Sneed and Folk (1958) methods were used in this study. Both methods are based on the combination of three orthogonal axial lengths: L, I, and S (respectively, the lengths of the longest, intermediate, and shortest axes of individual clasts), so that $L \geq I \geq S$. The Zingg (1935) method allows for differentiating four classes of shapes (spheroids, blades, discoids, and rods) through a Cartesian coordinate system using the indices S/I and I/L. The ratio S/I discriminates blades and discs (low values) from rods and spheres (high values), while the ratio I/L separates blades and rods (low values) from discs and spheres (high values).

To exhibit shape-sorting properties, the disc:rod index from Sneed and Folk ($DRI = (L - I)/(L - S)$) was applied. These authors separated oblate (discoid) shapes from prolate (blade and rod) shapes and suggested the combination of DRI with the ratio S/L to produce three end members of shape (disc, rod, and sphere) in a triangular diagram. In addition, we used the effective settling sphericity index of Sneed and Folk (1958) as the maximum projection sphericity of a particle (ψ_p), which is given by the equation $\psi_p = (S^2/L{\cdot}I)^{1/3}$.

Void ratios were estimated from the grain size distribution, sphericity and roundness indices, and slenderness ratio [38]. These relationships have already been proven by various authors [39,40]. A wider grain size distribution allows the particles to compact more densely, as the smaller particles fill in the gaps between their larger neighbors. Particles that are more angular pack less tightly, as sharp corners pull them apart. Increasing the slenderness of the particles can decrease the density as the thin particles close the gaps between the grains and create large open voids. To calculate hydraulic conductivity (HK) ($m.s^{-1}$) we used the Kozeny-Carman equation from effective porosity and characteristic grain sizes, according to its most simplified form [35].

3.3. Statistical Relationship between Texture Parameters and Electrical Resistivity

The Past program was used for the statistical analysis [41]. Shapiro-Wilk W was applied for testing normal distribution. Univariate analysis included examining Pearson's correlation between sediment texture from borehole logs and the values of resistivity from ETR. The resistivity values correspond to the average of ETR records for each depth section with a homogeneous structure and materials extracted from the boreholes. The tests took two levels of significance for the p value ($p < 0.05$ ** and $p < 0.1$ *).

The variables with the highest correlation were selected to develop linear regression models using the least squares method. The values of the texture analysis as dependent variables and the resistivity data as explanatory variables were applied. To further explore the relationship between these variables, the results were adjusted to polynomial models, which served to minimise the variance of the unbiased estimators of the coefficients.

4. Results and Discussion

4.1. Changes in Sediment Texture from Borehole Sample Datasets

Table 2 shows the characteristic particle sizes and sorting indices obtained for all clastic sediment samples that were extracted from the boreholes. According to these data, different patterns in thickness and sediment texture were found in the alluvial channel-fill: the pattern located in the active channel, characterised by a progressive decrease in the median grain size, and an increase in the sorting index with depth, passing from medium-sized gravels at the surface level to sands and small gravels at the bottom (BH3); the pattern defined in the lateral bars (BH1 and BH4) by a single layer 1 to 1.5 m thick composed of medium and coarse gravels and moderate heterometry ($2.1 < \sigma < 2.4$), directly lying on a softer rocky substrate; and a third pattern in the edge and tail part of a central bar (BH2), where a subsurface level of coarse material is prevalent, mainly gravels with a size dispersion index somewhat smaller than in the previous cases, but noting an abrupt change in texture in depth (a layer with abundant sand 1 m thick and a deeper and slightly thinner mixed layer of sands, gravels, and pebbles). In general, except in the BH2 case, the subsurface detritical layer has a homogeneous thickness, between 1 and 1.5 m, and consists of coarser materials than deep deposits.

Table 2. Representative grain sizes and sorting estimated for clastic sediments in borehole logs. Rambla de la Azohía.

		Grain Sizes				
	Depth (m)	**D_{10} (mm)**	**D_{20} (mm)**	**D_{50} (mm)**	**D_{84} (mm)**	**σ**
BH1	0.0–1.1	0.4	2.1	15	36	2.37
	1.1–3.0	0.01	0.02	0.04	0.06	0.08
BH2	0.0–1.4	0.4	2.5	12	27	2.09
	1.4–2.3	0.2	1.1	1.9	15	2.21
	2.3–3.0	0.5	2.6	17	32	2.22
BH3	0.0–1.2	0.4	3.3	10	17	1.86
	1.2–1.6	0.3	1.9	8	18	2.08
	1.6–2.5	0.4	1.4	6	26	2.20
	2.5–3.0	0.1	0.5	4	12	2.73
BH4	0.0–1.5	0.5	3.2	17	27	2.11

BH is the borehole code, D_{10} and D_{20} are the effective grain sizes corresponding to 10% and 20% of the total sample weight (mm), D_{50} is the median grain size (mm), D_{84} is the particle size for 84% of the sample weight (mm), and σ is the inclusive graphic standard deviation (sorting) after Folk (1974).

Average values of the Zingg and the Sneed and Folk shape and sphericity indices for each depth interval are shown in Table 3. A common feature in all borehole logs is the decrease in the percentage of rod particles as depth increases. This fact may have conditioned the void ratio and the effective porosity in a similar proportion, also affecting the obtained electrical resistivity values. It is also worth noting the progressive decrease in the I/L ratio in the discoid or oblate class obtained according to the depth for the BH3 borehole. The highest percentages reached in the subsurface layer correspond to the discoid/oblate and rod classes (>80% between both classes), which tend to increase the degree of porosity at this level. On the other hand, the disc-rod index (DRI) and the effective settling sphericity (ψ_p) do not seem to be decisive for differentiating sediment textures, since they range between 0.38 and 0.51 in the first case and between 0.60 and 0.69 in the second.

Table 3. Average values of the Zingg and the Sneed and Folk shape and sphericity indices, estimated for gravels and pebbles in borehole logs. Rambla de la Azohía.

	Depth (m)	Zingg Shape Classes												Shape—Sphericity (Sneed & Folk, 1958)		
		Discoid or Oblate			Equid.—Spheroid			Blade			Rod					
		%	I/L	S/L	%	I/L	S/L	%	I/L	S/L	%	I/L	S/L	S/L	DRI	ψ_P
BH1	0.0–1.1	57.1	0.81	0.55	0.0	-	-	14.3	0.45	0.60	28.6	0.58	0.75	0.42	0.51	0.63
	1.1–3.0	Loamy substrate														
BH2	0.0–1.4	50.0	0.82	0.50	16.7	0.80	0.71	0.0	-	-	33.3	0.47	0.71	0.41	0.49	0.62
	1.4–2.3	50.0	0.72	0.50	16.7	0.70	0.89	16.7	0.60	0.50	16.7	0.63	0.68	0.40	0.55	0.62
	2.3–3.0	50.0	0.80	0.54	0.0	-	-	50.0	0.61	0.59	0.0	-	-	0.39	0.47	0.60
BH3	0.0–1.2	33.3	0.93	0.46	0.0	-	-	16.7	0.59	0.58	50.0	0.59	0.74	0.42	0.51	0.63
	1.2–1.6	50.0	0.89	0.54	50.0	0.76	0.72	0.0	-	-	0.0	-	-	0.51	0.38	0.69
	1.6–2.5	100.0	0.77	0.60	0.0	-	-	0.0	-	-	0.0	-	-	0.46	0.42	0.65
	2.5–3.0	50.0	0.73	0.49	33.3	0.74	0.75	0.0	-	-	16.7	0.62	0.81	0.44	0.54	0.65
BH4	0.0–1.5	57.1	0.76	0.60	14.3	0.83	0.84	28.6	0.63	0.55	0.0	-	-	0.45	0.49	0.65

S, I and L are measurements of the short, intermediate, and long orthogonal axes of clasts, made from selected sieve intervals that represent the spread in the size distribution of the natural gravel and pebbles; DRI is the disc–rod index $(L - I)/(L - S)$; ψ_P is the effective settling sphericity.

4.2. Electrical Resistivity Tomography 2D Survey

As a general rule, the penetration depth is approximately 15–25% of the longest array length for any four-electrode array [42]. Resistivity results of this study present higher values owing to the sort and material type as Grygar et al. [43] and Chaudhuri et al. [44] have alluded in their study. The ERT4 profile is the largest profile of the whole study that reaches around 6 m of investigation depth. This profile identifies the internal structure of the gravel-bed channel. BH4 was made as to coincide with the ERT4 profile approximately in the central part of the profile, as expected [6,7]. The depth at which the marl substrate is encountered is 1.1 m. Additionally, profile ERT4 intersects with profiles ERT1, ERT2, and ERT3 at 11 m, maintaining a uniform distribution of the position of the marl substrate (Figure 3).

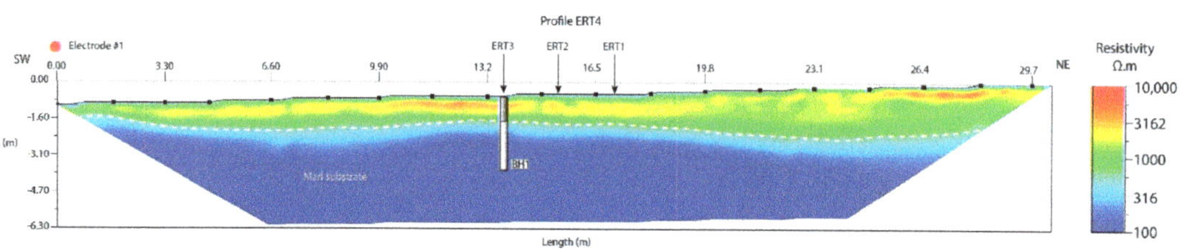

Figure 3. Perpendicular ERT profile ERT4. The other three profiles intersect vertical with arrows to mark the position of ERT1-ERT3.

At the top, an alluvial channel-fill is identified with high values of resistivity above ≈600 Ω.m, while at the bottom, the resistivity progressively decreases to low values (<300 Ω.m), indicating the presence of the marl substrate. The white dashed line marks the border between them, see Figure 3.

The ERT4 profile provides a general overview of the channel. The profiles ERT1-ERT3 were limited by the channel width, reaching a depth of investigation of only ≈3 m. Figure 4 shows the resulting 2D geoelectrical sections of ERT1, ERT2, and ERT3 profiles. After the eighth iteration, the ERT1 profile yielded a section with a root mean square (RMS) of 9.47% and 1.05 of L2. In the case of the ERT2 profile, the RMS is 9.51% and L2 = 1.03, and finally, the ERT3 profile presents RMS = 10.21% and L2 = 1.09.

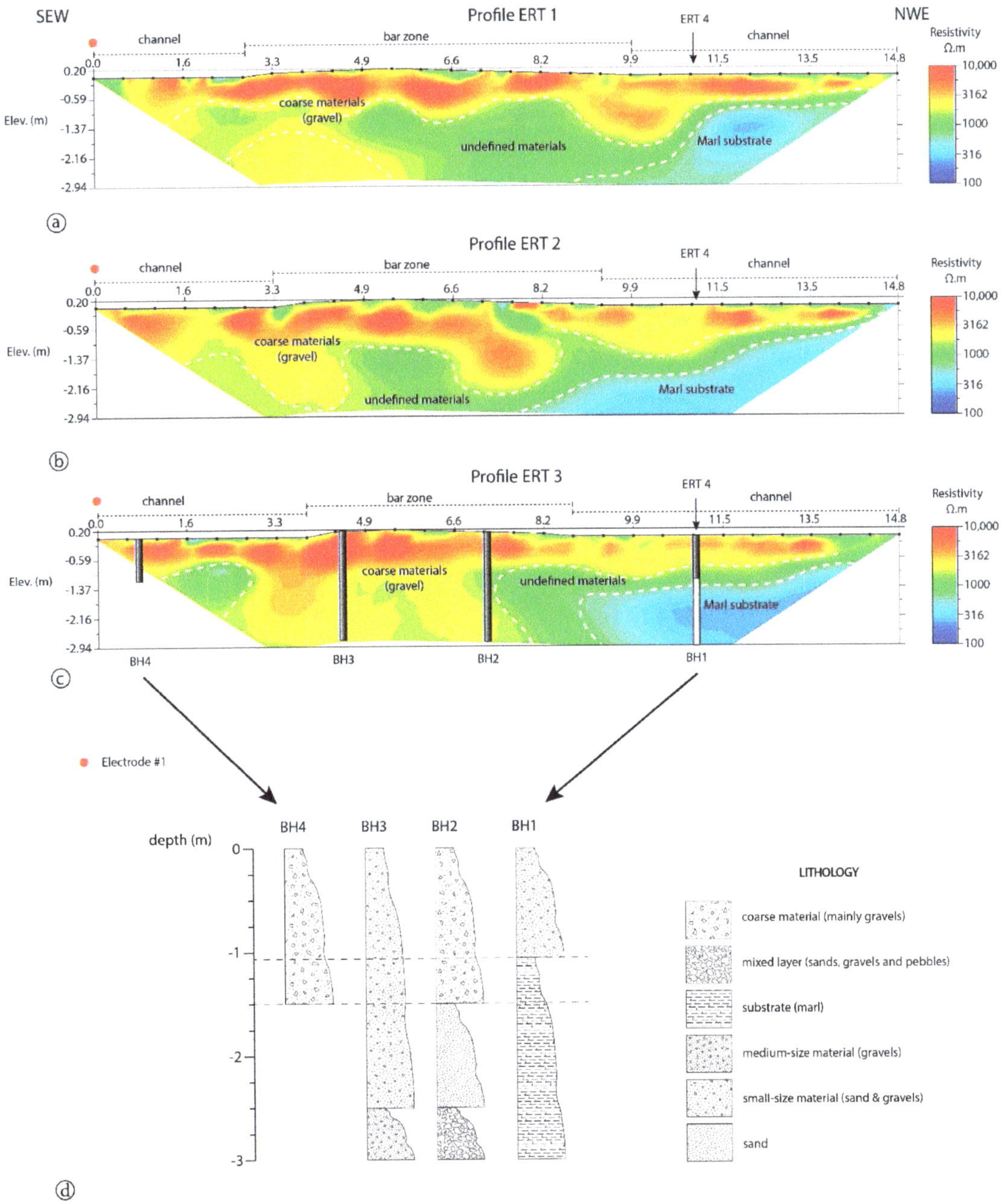

Figure 4. (**a**–**c**) Parallel ERT profiles (ERT1-ERT3). ERT4 intersects all profiles. BH1-BH4 are the boreholes drilled in the study area. (**d**). Lithology from BH1-BH4.

Besides, the profiles ERT1–ERT3 identified the marl substrate featured with lower resistivity values (blue-cyan in the colour bar), which confirms the results obtained from the ERT4 profile. The 2D ERT study reproduced the physical structure of the channel, it presented a central bar and two lateral channels. ERT1 was in the upstream while

ERT3 was in the downstream direction. This distribution allowed us to identify the bar geometry, which is wider in ERT1 but narrower in ERT3. The bar presents a triangular form, indicating that the bar decreases in the downstream direction. Four mechanical boreholes were drilled on ERT3 (Figure 4).

Figure 4 presents a horizontal layer with a relatively constant thickness of about ≈1 m which is associated with a high resistivity (>2000 Ω.m). The results from boreholes support ERT results and link this layer to the pattern located in the active channel composed of medium and coarse gravels. The third pattern is related to BH2. In this pattern exists a top layer composed of coarser materials (gravel), corresponding to resistivities > 5000 Ω.m (ERT3) and a second layer of sand layer 1 m thick. The sand layer does not appear in the profile ERT 3 although it appears in profile ERT2 with resistivities > 3000 Ω.m. It was not possible to gather a sample from the material featured in ERT with a green colour (1000–2000 Ω.m); therefore, we cannot define the exact material type (Figure 4).

The results of the 2D study fit coherently with the texture datasets provided by the borehole logs and what is expected for the study area. In general, at the top, a detrital layer, 1.2 m thick, showed especially high ERT values. However, this did not always match the coarser grain sizes. BH3 resistivities between 4000 and 9000 Ω.m were recorded in medium gravels with a significant effective grain size (D_{20}) (3.3 mm), due to matrix composed of clay and sand, which decreased the porosity estimates (ϕ = 0.27) and hydraulic conductivity (CH $\leq 6\ 10^{-4}$ m.s^{-1}). This high relationship between ERT and D_{20} was also found in BH2 and BH4, where the highest resistivity (5000 to 8000 Ω.m) occurred in the subsurface deposits with greater D_{20} grain sizes (2.5 to 3.2 mm), and lower HC ($<7 \times 10^{-4}$ m.s^{-1}). In BH1, located on the right part of the channel, the upper sedimentary layer (1 m thick) showed moderate ERT values, associated with the presence of medium gravels and higher porosity (ϕ = 0.35). The central part a longitudinal bar has been developed, whose recent evolution reflects a sedimentary pattern typical of several ephemeral gravel-bed streams, which are subjected to the effects of climate change: a progressive decrease in particle size with depth and strong surface armouring. Conversely, the sorting index increased towards the deeper alluvial layers.

4.3. Electrical Resistivity Tomography 3D Modelling

4.3.1. Statistical Relationships

Some significant relationships between physical properties and electrical resistivity tomography were found. Table 4 summarises the results. Four variables showed the most significant correlation coefficients. One describes the grain size and the others are linked to the particle shape. Among all, the most significant is D_{20} with r = 0.67.

Table 4. Pearson's correlation results.

	Grain Sizes					Zingg Shape Classes											
	D_{10} (mm)	D_{20} (mm)	D_{50} (mm)	D_{84} (mm)	σ	Discoid or Oblate			Equidimensional—Spheroid			Blade			Rod		
						%	I/L	S/L	%	I/L	S/L	%	I/L	S/L	%	I/L	S/L
p	0.48	0.05	0.31	0.52	0.43	0.09	0.09	0.14	0.64	0.11	0.96	0.69	0.66	0.88	0.07	0.66	0.38
r	0.27	0.67	0.36	0.23	0.28	−0.60	0.60	−0.53	−0.18	−0.79	−0.03	0.15	0.27	−0.09	0.63	−0.27	−0.51
n	9	9	10	10	10	9	9	9	9	5	5	9	5	5	9	5	5

p: *p* value; r: Pearson's coefficient correlation; n: sample size.

Following these correlations, we perform a regression analysis. Figure 5 shows the four regression models obtained. Rod (%), Discoid (I/L), and D_{20} (mm) present a direct relationship with the resistivity. While one variable increases, the other variable has the same behaviour. Discoid (%) presents an indirect relationship with resistivity, which means that while the resistivity rises, the discoid variable reduces.

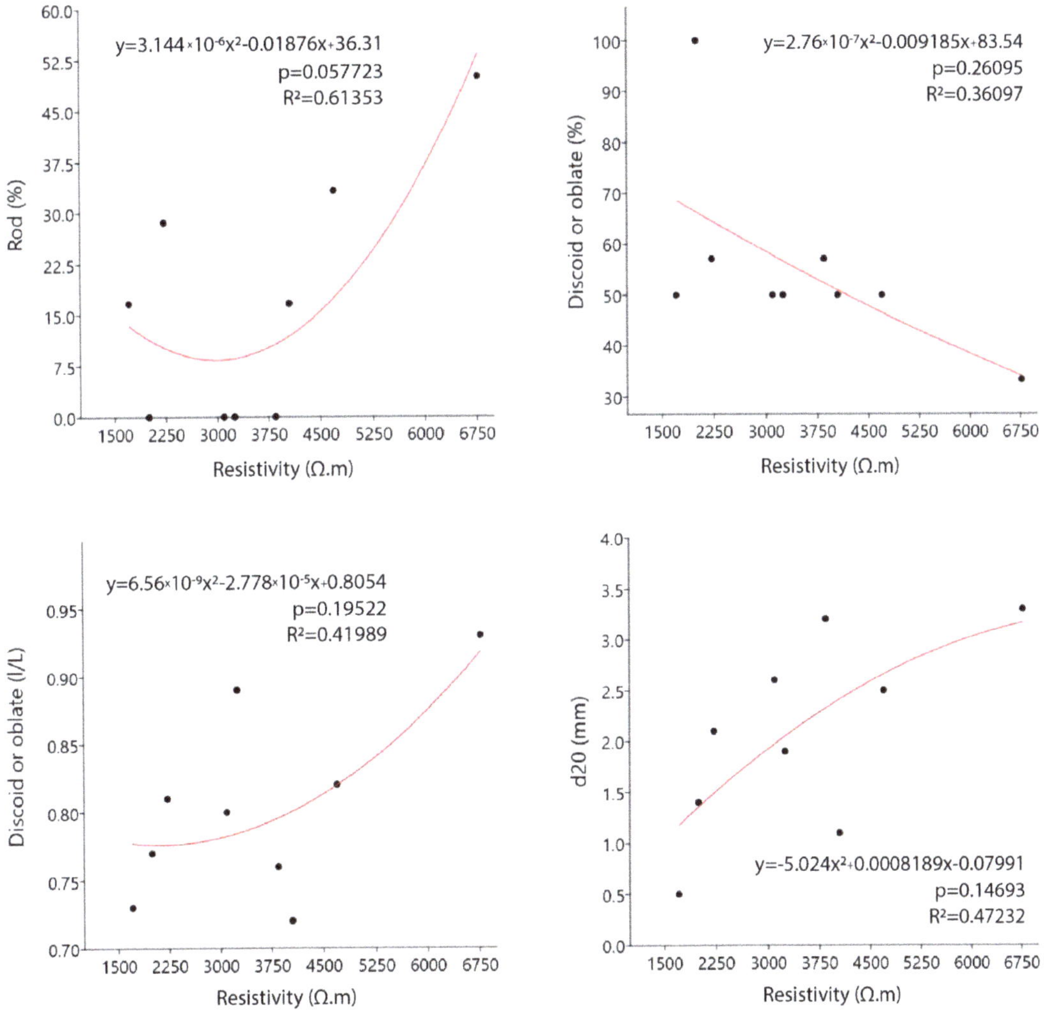

Figure 5. Regression models showing the relationships between resistivity and particle-shape variables.

4.3.2. 3D Model

As direct measures for supporting the electrical resistivity tomography study, four boreholes were drilled on the ERT3 profile. These were distributed uniformly along the profile. Table 5 contains the summary of the study results. Samples at different depths of each borehole were employed to perform the analysis. A 3D model is a proven tool that contributes significantly to identifying different subsurface structures used in several studies [45,46].

Table 5. Resistivity results from the ERT study.

	Resistivity Values in Ω.m					
	Depth	Max	Min	Mean	SD	COV
	m					%
BH1	0.0–1.1	2543.4	1590.3	2187.2	252.9	11.6
	1.1–3.0	1427.3	318.7	595.9	253.6	42.6
BH2	0.0–1.4	6923.6	3370.8	4695.5	1303.5	27.8
	1.4–2.3	4304.8	3558.7	4023.4	211.6	5.3
	2.3–3.0	3558.7	2919.3	3160.4	228.4	7.2
BH3	0.0–1.2	8599.2	3744.0	6623.5	1447.4	21.9
	1.2–1.6	3744.0	2694.3	3180.3	309.8	9.7
	1.6–2.5	2694.3	1716.1	1984.3	276.4	13.9
	2.5–3.0	1716.1	1706.3	1710.0	4.3	0.3
BH4	0.0–1.5	7183.5	1832.7	3843.7	1665.2	43.3

BH: Boreholes, SD: Standard deviation, COV: Coefficient of variation.

The 3D model computed with the electrical resistivity tomography data identified the same changes in sediment texture (patterns) as those found in the grain size study [47]. The colour bar of the model was calibrated with 1000 Ω.m as the step value. This value allowed for the distinguishing of the underground formations previously found in the 2D study, and the visualising of the extent of them simultaneously. Figure 6 shows the different patterns, which varied in accordance with the increase/decrease in the resistivity.

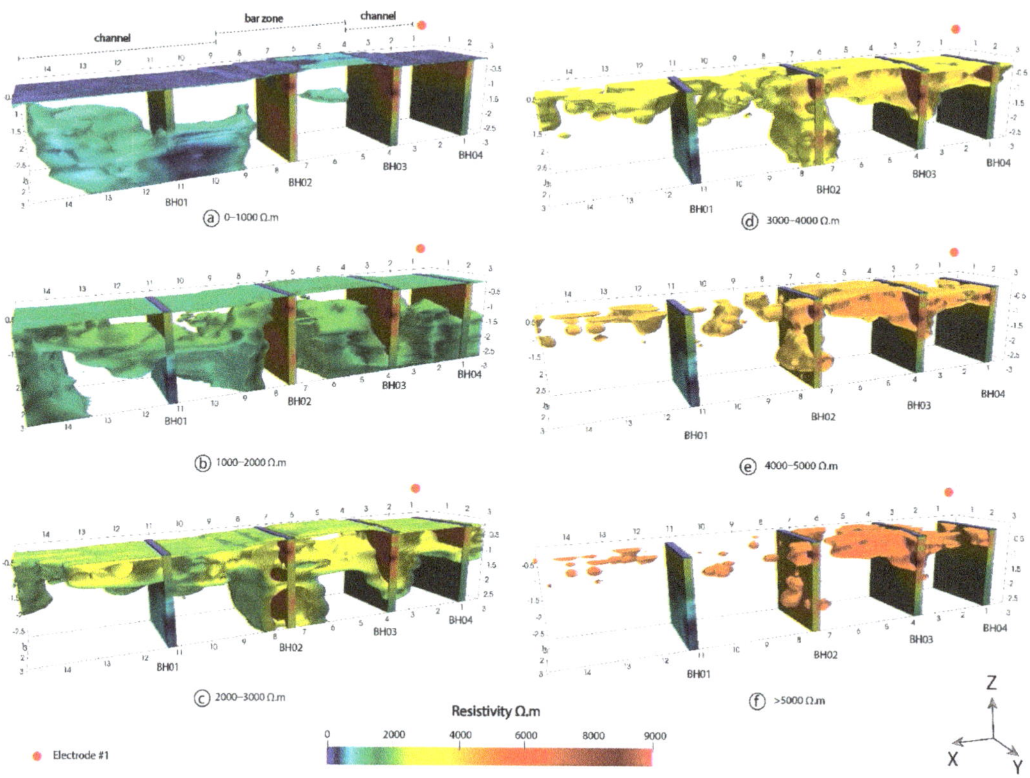

Figure 6. 3D model. (**a**–**f**) illustrates the different layers detected by the electrical resistivity tomography survey.

Figure 6 shows the 3D model; this model has four vertical sections in the same place where the boreholes were drilled, aiming to facilitate the data interpreting process and visualising how the material evolves in that space. Figure 6a–f presents the different layers and body formations continuously. Notice that while the resistivity values rise, materials tend to group in the central site of the model, which corresponds to bar place.

Figure 6a represents to the lowest resistivity values (0–1000 Ω.m) of the whole study. These values belong to the softer rocky substrate, which is in line with BH1 results, and show that the model is a realistic reproduction of the study area. Figure 6b presents a new layer featured with the green colour (1000–2000 Ω.m), which is spread in the whole model. This layer, representing medium gravel with little clay matrix, cut the borehole BH4 at 1.5 m deep. Since the samples collected from BH4 only reached to a depth of 1.5 m, this pattern was not contemplated in the granulometric study; however, the ERT detected its presence.

Figure 6c–f, with values of resistivity >2000 Ω.m, represent the layer 1 to 1.5 m thick, which is common in all boreholes, except in BH2, composed mainly of medium and coarse gravels. The layer disposition in the 3D model adopts an onion-like shape, finding the least resistivity values in the external part and the most resistant in the core. Values of resistivity ranging from 2000 to 3000 Ω.m (Figure 6c) are linked to the green-yellow colour, forming a layer in the model that constrains in the lateral direction of the bar by 6–8 m, acquiring a shell form at the bottom of the model.

The inner part of the onion-like structure is represented in Figure 6d–f, which is focused on BH2 (7.5 m). The upper part of the borehole (Figure 6f) corresponds to coarse materials (gravel and pebbles) linked to resistivity values above 5000 Ω.m. An intermediate layer (Figure 6d,e), 1 m thick, composed of gravel with abundant sand, reduces the range of resistivity at the middle stretch of BH2 (3000–5000 Ω.m).

In general, the 3D model detected in the central channel part an upper subsurface layer of 1–1.2 m thick, with resistivities above 4500 Ω.m, which corresponded to the coarsest materials (gravel and pebbles) and gravels with a supported matrix. In addition, resistivity reduces according to depth (Table 5), owing to a progressive decrease in the median grain size and an increase in the sorting index with depth. The lateral parts, close to the banks, have undergone a different geomorphological evolution: the right side currently constitutes the active channel of low waters, with a uniform bed composed of medium gravels, which, during much of the filling period, received finer-sized materials (small gravels and sands). The resistivities in this case decreased from 4000 Ω.m at the top to 500 Ω.m at the limit with the loamy substrate. On the other hand, the left part shows a higher concentration of thicker sediments due to the presence of a lateral bar still under development today, which is in accordance with the measured ERT values in the 3D model (>4500 Ω.m).

5. Conclusions

The physical characterisation of the alluvial deposit was successfully completed. The resistivity trend is high, owing to the presence of air and the loose nature of the materials. The 3D ERT models obtained have allowed for the estimation of both the vertical and horizontal variations of the thickness, and the characterisation of the different materials. Borehole logs are a punctual sample, whereas electrical resistivity tomography allows for inferring the 3D information of the deposit.

The combination of electrical resistivity tomography, 2D and 3D, with some borehole logs have demonstrated that they are a useful methodology for assessing alluvial deposits. Consequently, the ERT results coupled with borehole data suggest that from the channel entrenchment in the Miocene marl substrate, different pulses of vertical sedimentary accretion were produced, denoting a general trend of increasing grain size (coarsening upwards) towards the top of the sedimentary sequence.

Author Contributions: Conceptualisation, C.C.-G. and M.A.M.-S.; methodology, P.P.-C., M.D.V.-M., C.C.-G., and M.A.M.-S.; software, M.A.M.-S.; validation, P.M.-P., P.P.-C., M.D.V.-M., C.C.-G., and M.A.M.-S.; formal analysis, C.C.-G. and M.A.M.-S.; investigation, P.M.-P., P.P.-C., M.D.V.-M., C.C.-G., and M.A.M.-S.; resources, C.C.-G.; data curation, P.P.-C., M.D.V.-M., C.C.-G., and M.A.M.-S.; writing—original draft preparation, M.D.V.-M., C.C.-G., and M.A.M.-S.; writing—review and editing, P.M.-P., P.P.-C., M.D.V.-M., C.C.-G., and M.A.M.-S.; visualisation, P.M.-P., P.P.-C., M.D.V.-M., C.C.-G., and M.A.M.-S.; supervision, P.M.-P., P.P.-C., M.D.V.-M., C.C.-G., and M.A.M.-S.; project administration, C.C.-G.; funding acquisition, C.C.-G. All authors have read and agreed to the published version of the manuscript.

Funding: This research was funded by ERDF/Spanish Ministry of Science, Innovation and Universities—State Research Agency/Project CGL2017-84625-C2-1-R (CCAMICEM); State Program for Research, Development and Innovation focused on the Challenges of Society.

Institutional Review Board Statement: Not applicable.

Informed Consent Statement: Not applicable.

Data Availability Statement: Not applicable.

Acknowledgments: The authors would also like to extend their thanks to the Segura River Hydrographic Confederation Centre (SHC), Government of Spain, for its collaboration.

Conflicts of Interest: The authors declare no conflict of interest.

References

1. Bull, W.B. Discontinuous ephemeral streams. *Geomorphology* **1997**, *19*, 227–276. [CrossRef]
2. Segura-Beltrán, F.; Sanchis-Ibor, C. Assessment of channel changes in a Mediterranean ephemeral stream since the early twentieth century. The Rambla de Cervera, eastern Spain. *Geomorphology* **2013**, *201*, 199–214. [CrossRef]
3. Norman, L.M.; Sankey, J.B.; Dean, D.; Caster, J.; DeLong, S.; DeLong, W.; Pelletier, J.D. Quantifying geomorphic change at ephemeral stream restoration sites using a coupled-model approach. *Geomorphology* **2017**, *283*, 1–16. [CrossRef]
4. Vásconez-Maza, M.D.; Martínez-Segura, M.A.; Bueso, M.C.; Faz, Á.; García-Nieto, M.C.; Gabarrón, M.; Acosta, J.A. Predicting spatial distribution of heavy metals in an abandoned phosphogypsum pond combining geochemistry, electrical resistivity tomography and statistical methods. *J. Hazard. Mater.* **2019**, *374*, 392–400. [CrossRef] [PubMed]
5. Martínez-Segura, M.A.; Vásconez-Maza, M.D.; Faz, Á.; Martínez-Martínez, S.; Gabarrón, M.; Acosta, J.A. Assessment of the environmental impact of an agricultural area near a coastal lagoon using geophysical and geochemical techniques. In *Recent Advances in Geophysics*; Nova Science Publishers, Inc.: Hauppauge, NY, USA, 2019; ISBN 978-1-53616-207-3.
6. Uhlemann, S.; Kuras, O.; Richards, L.A.; Naden, E.; Polya, D.A. Electrical resistivity tomography determines the spatial distribution of clay layer thickness and aquifer vulnerability, Kandal Province, Cambodia. *J. Asian Earth Sci.* **2017**, *147*, 402–414. [CrossRef]
7. Bábek, O.; Sedláček, J.; Novák, A.; Létal, A. Electrical resistivity imaging of anastomosing river subsurface stratigraphy and possible controls of fluvial style change in a graben-like basin, Czech Republic. *Geomorphology* **2018**. [CrossRef]
8. Chambers, J.; Ogilvy, R.; Kuras, O.; Cripps, J.; Meldrum, P. 3D electrical imaging of known targets at a controlled environmental test site. *Environ. Geol.* **2002**, *41*, 690–704. [CrossRef]
9. Ogilvy, R.D.; Meldrum, P.I.; Kuras, O.; Wilkinson, P.B.; Chambers, J.E.; Sen, M.; Pulido-Bosch, A.; Gisbert, J.; Jorreto, S.; Frances, I.; et al. Automated monitoring of coastal aquifers with electrical resistivity tomography. *Near Surf. Geophys.* **2009**, *7*, 367–376. [CrossRef]
10. Ortega, J.A.; Razola, L.; Garzón, G. Recent human impacts and change in dynamics and morphology of ephemeral rivers. *Nat. Hazards Earth Syst. Sci.* **2014**, *14*, 713–730. [CrossRef]
11. Conesa-García, C.; Puig-Mengual, C.; Riquelme, A.; Tomás, R.; Martínez-Capel, F.; García-Lorenzo, R.; Pastor, J.L.; Pérez-Cutillas, P.; Cano Gonzalez, M. Combining sfm photogrammetry and terrestrial laser scanning to assess event-scale sediment budgets along a gravel-bed ephemeral stream. *Remote Sens.* **2020**, *12*, 3624. [CrossRef]
12. Everett, M.E. *Near-Surface Applied Geophysics*; Cambridge University Press: Cambridge, UK, 2013; ISBN 9781107018778.
13. Vásconez-Maza, M.D.; Martínez-Pagán, P.; Aktarakçi, H.; García-Nieto, M.C.; Martínez-Segura, M.A. Enhancing Electrical Contact with a Commercial Polymer for Electrical Resistivity Tomography on Archaeological Sites: A Case Study. *Materials* **2020**, *13*, 5012. [CrossRef] [PubMed]
14. AGI EarthImager 3D Resistivity Software. Available online: https://www.agiusa.com/agi-earthimager-3d (accessed on 2 February 2021).
15. Loke, M.H. Tutorial:2-D and 3-D Electrical Imaging Surveys. Available online: http://citeseerx.ist.psu.edu/viewdoc/download?doi=10.1.1.454.4831&rep=rep1&type=pdf (accessed on 5 March 2019).
16. Kasprzak, M. High-resolution electrical resistivity tomography applied to patterned ground, Wedel Jarlsberg Land, south-west Spitsbergen. *Polar Res.* **2015**, *34*, 1–13. [CrossRef]

17. Greggio, N.; Giambastiani, B.M.S.; Balugani, E.; Amaini, C.; Antonellini, M. High-resolution electrical resistivity tomography (ERT) to characterize the spatial extension of freshwater lenses in a salinized coastal aquifer. *Water* **2018**, *10*, 1067. [CrossRef]
18. Rubio-Melendi, D.; Gonzalez-Quirós, A.; Roberts, D.; García García, M.D.C.; Caunedo Domínguez, A.; Pringle, J.K.; Fernández-Álvarez, J.P. GPR and ERT detection and characterization of a mass burial, Spanish Civil War, Northern Spain. *Forensic Sci. Int.* **2018**, *287*, e1–e9. [CrossRef] [PubMed]
19. Dépret, T.; Virmoux, C.; Gautier, E.; Piégay, H.; Doncheva, M.; Plaisant, B.; Ghamgui, S.; Mesmin, E.; Saulnier-Copard, S.; de Milleville, L.; et al. Lowland gravel-bed river recovery through former mining reaches, the key role of sand. *Geomorphology* **2021**, *373*, 107493. [CrossRef]
20. Blackburn, J.; Comte, J.-C.; Foster, G.; Gibbins, C. Hydrogeological controls on the flow regime of an ephemeral temperate stream flowing across an alluvial fan. *J. Hydrol.* **2021**, *595*, 125994. [CrossRef]
21. Rey, J.; Martínez, J.; Hidalgo, M.C. Investigating fluvial features with electrical resistivity imaging and ground-penetrating radar: The Guadalquivir River terrace (Jaen, Southern Spain). *Sediment. Geol.* **2013**, *295*, 27–37. [CrossRef]
22. Loke, M.H.; Chambers, J.E.; Rucker, D.F.; Kuras, O.; Wilkinson, P.B. Recent developments in the direct-current geoelectrical imaging method. *J. Appl. Geophys.* **2013**, 135–156. [CrossRef]
23. AGI Dipole-Dipole Array. Available online: https://www.agiusa.com/dipole-dipole-array (accessed on 16 July 2020).
24. AGI SuperSting Wi-Fi. Available online: https://www.agiusa.com/supersting-wifi (accessed on 16 July 2020).
25. Loke, M.H.; Barker, R.D. Rapid least-squares inversion of apparent resistivity pseudosections by a quasi-Newton method1. *Geophys. Prospect.* **1996**, *44*, 131–152. [CrossRef]
26. Chambers, J.E.; Wilkinson, P.B.; Wardrop, D.; Hameed, A.; Hill, I.; Jeffrey, C.; Loke, M.H.; Meldrum, P.I.; Kuras, O.; Cave, M.; et al. Bedrock detection beneath river terrace deposits using three-dimensional electrical resistivity tomography. *Geomorphology* **2012**, *177–178*, 17–25. [CrossRef]
27. SIAM Ficha de Estaciones. Available online: http://siam.imida.es/apex/f?p=101:1000:6322900036254158::NO::: (accessed on 17 March 2021).
28. Bear, J. *Dynamics of Fluids in Porous Media*; Dover Publications, Inc.: New York, NY, USA, 2013; ISBN 978-0-486-65675-5.
29. Salisbury, J.W.; Eastes, J.W. The effect of particle size and porosity on spectral contrast in the mid-infrared. *Icarus* **1985**, *64*, 586–588. [CrossRef]
30. Jin, G.; Torres-Verdín, C.; Lan, C. Pore-level study of grain-shape effects on petrophysical properties of porous media. In Proceedings of the SPWLA 50th Annual Logging Symposium 2009, The Woodlands, TX, USA, 21–24 June 2009.
31. Tucker, M. *Sedimentary Petrology: An Introduction to the Origin of Sedimentary Rocks*, 3rd ed.; Blackwell Science Ltd.: Oxford, UK, 2001; ISBN 978-0-632-05735-1.
32. Beard, D.C.; Weyl, P.K. Influence of Texture on Porosity and Permeability of Unconsolidated Sand. *Am. Assoc. Pet. Geol. Bull.* **1973**, *57*. [CrossRef]
33. Vukovic, M.; Soro, A. *Determination of Hydraulic Conductivity of Porous Media from Grain-Size Composition*; Water Resources Publications: Littleton, CO, USA, 1992; ISBN 9780918334770.
34. Cheng, C.; Chen, X. Evaluation of methods for determination of hydraulic properties in an aquifer-aquitard system hydrologically connected to a river. *Hydrogeol. J.* **2007**, *15*, 669–678. [CrossRef]
35. Odong, J. Evaluation of Empirical Formulae for Determination of Hydraulic Conductivity based on Grain-Size Analysis. *J. Am. Sci.* **2007**, *3*, 54–60.
36. Koch, K.; Kemna, A.; Irving, J.; Holliger, K. Impact of changes in grain size and pore space on the hydraulic conductivity and spectral induced polarization response of sand. *Hydrol. Earth Syst. Sci.* **2011**, *15*, 1785–1794. [CrossRef]
37. Urumovi, K.; Urumović, K., Sr. The effective porosity and grain size relations in permeability functions. *Hydrol. Earth Syst. Sci. Discuss.* **2014**, *11*, 6675–6714. [CrossRef]
38. Cho, G.-C.; Dodds, J.; Santamarina, J.C. Particle Shape Effects on Packing Density, Stiffness, and Strength: Natural and Crushed Sands. *J. Geotech. Geoenvironmental Eng.* **2006**, *132*, 591–602. [CrossRef]
39. Dodds, J.; Santamarina, C.; Paul Mayne Glenn Rix, A. *Particle Shape and Stiffness-Effects on Soil Behavior*; Georgia Institute of Technology: Atlanta, GA, USA, 2003.
40. Rousé, P.C.; Fannin, R.J.; Shuttle, D.A. Influence of roundness on the void ratio and strength of uniform sand. *Géotechnique* **2008**, *58*, 227–231. [CrossRef]
41. Hammer, Ø.; Harper, D.A.T.; Ryan, P.D. PAST: Paleontological Statistics Software Package for Education and Data Analysis. *Palaeontol. Electron.* **2001**, *4*, 1–9.
42. AGI Quick Tip: Depth of Investigation for ERI Surveys. Available online: https://www.agiusa.com/blog/quick-tip-depth-investigation-eri-surveys (accessed on 28 February 2021).
43. Matys Grygar, T.; Elznicová, J.; Tůmová, S.; Faměra, M.; Balogh, M.; Kiss, T. Floodplain architecture of an actively meandering river (the Ploučnice River, the Czech Republic) as revealed by the distribution of pollution and electrical resistivity tomography. *Geomorphology* **2016**. [CrossRef]
44. Chaudhuri, A.; Sekhar, M.; Descloitres, M.; Godderis, Y.; Ruiz, L.; Braun, J.J. Constraining complex aquifer geometry with geophysics (2-D ERT and MRS measurements) for stochastic modelling of groundwater flow. *J. Appl. Geophys.* **2013**, *98*, 288–297. [CrossRef]

45. Zeng, R.Q.; Meng, X.M.; Zhang, F.Y.; Wang, S.Y.; Cui, Z.J.; Zhang, M.S.; Zhang, Y.; Chen, G. Characterizing hydrological processes on loess slopes using electrical resistivity tomography—A case study of the Heifangtai Terrace, Northwest China. *J. Hydrol.* **2016**, *541*, 742–753. [CrossRef]
46. Argote-Espino, D.L.; López-García, P.A.; Tejero-Andrade, A. 3D-ERT geophysical prospecting for the investigation of two terraces of an archaeological site northeast of Tlaxcala state, Mexico. *J. Archaeol. Sci. Rep.* **2016**, *8*, 406–415. [CrossRef]
47. Chambers, J.E.; Wilkinson, P.B.; Penn, S.; Meldrum, P.I.; Kuras, O.; Loke, M.H.; Gunn, D.A. River terrace sand and gravel deposit reserve estimation using three-dimensional electrical resistivity tomography for bedrock surface detection. *J. Appl. Geophys.* **2013**, *93*, 25–32. [CrossRef]

applied sciences

MDPI

Article

Combining of MASW and GPR Imaging and Hydrogeological Surveys for the Groundwater Resource Evaluation in a Coastal Urban Area in Southern Spain

Francisco Javier Alcalá [1,2,*], Pedro Martínez-Pagán [3], Maria Catarina Paz [4], Manuel Navarro [5], Jaruselsky Pérez-Cuevas [6] and Francisco Domingo [7]

1 Instituto Geológico y Minero de España, 28003 Madrid, Spain
2 Facultad de Ingeniería, Instituto de Ciencias Químicas Aplicadas, Universidad Autónoma de Chile, 7500138 Santiago, Chile
3 Department of Mining and Civil Engineering, Universidad Politécnica de Cartagena, 30203 Cartagena, Spain; p.martinez@upct.es
4 CIQuiBio, Polytechnic Institute of Setúbal, Barreiro School of Technology, 2839-001 Lavradio, Portugal; catarina.paz@estbarreiro.ips.pt
5 Department of Chemistry and Physics, Universidad de Almería, 04120 Almería, Spain; mnavarro@ual.es
6 Engineering Sciences, Pontificia Universidad Católica Madre y Maestra, 2748 Santo Domingo, Dominican Republic; jaruselskyperez@pucmm.edu.do
7 Estación Experimental de Zonas Áridas, Consejo Superior de Investigaciones Científicas, 04120 Almería, Spain; poveda@eeza.csic.es
* Correspondence: fj.alcala@igme.es; Tel.: +34-91-349-5840

Citation: Alcalá, F.J.; Martínez-Pagán, P.; Paz, M.C.; Navarro, M.; Pérez-Cuevas, J.; Domingo, F. Combining of MASW and GPR Imaging and Hydrogeological Surveys for the Groundwater Resource Evaluation in a Coastal Urban Area in Southern Spain. *Appl. Sci.* **2021**, *11*, 3154. https://doi.org/10.3390/app11073154

Academic Editors: Micòl Mastrocicco, Francisco Javier Alcalá, Maria Catarina Paz, Pedro Martínez-Pagán and Fernando Monteiro Santos

Received: 27 February 2021
Accepted: 29 March 2021
Published: 1 April 2021

Featured Application: Combining of geophysical and hydrogeological surveys provides basic insights for the shallow groundwater resource evaluation in a coastal urban area.

Abstract: This paper conceptualizes and evaluates the groundwater resource in a coastal urban area hydrologically influenced by peri-urban irrigation agriculture. Adra town in southern Spain was the case study chosen to evaluate the groundwater resource contributed from the northern steep urban sector (NSUS) to the southern flat urban sector (SFUS), which belongs to the Adra River Delta Groundwater Body (ARDGB). The methodology included (1) geological and hydrogeological data compilation; (2) thirteen Multichannel Analysis of Surface Waves (MASW), and eight Ground Penetrating Radar (GPR) profiles to define shallow geological structures and some hydrogeological features; (3) hydrogeological surveys for aquifer hydraulic definition; (4) conceptualization of the hydrogeological functioning; and (5) the NSUS groundwater resource evaluation. All findings were integrated to prepare a 1:5000 scale hydrogeological map and cross-sections. Ten hydrogeological formations were defined, four of them (Paleozoic weathered bedrock, Pleistocene littoral facies, Holocene colluvial, and anthropogenic filling) in the NSUS contributing to the SFUS. The NSUS groundwater discharge and recharge are, respectively, around 0.28 Mm^3 $year^{-1}$ and 0.31 Mm^3 $year^{-1}$, and the actual groundwater storage is around 0.47 Mm^3. The groundwater renewability is high enough to guarantee a durable small exploitation for specific current and future urban water uses which can alleviate the pressure on the ARDGB.

Keywords: urban hydrogeology; hydrogeological map; multichannel analysis of surface waves; ground penetrating radar; aquifer geometry; groundwater resource evaluation; Adra town; Spain

1. Introduction

Water scarcity in many coastal drylands with limited surface water resources has propitiated high groundwater abstraction rates to supply the increasing urban, tourism, industrial, and agriculture demands [1]. This is the case in some coastal areas in Spain where aquifers play a critical role in sustaining the economy and the environment [2–4]. The combination of global climate forces underlying human pressures threatens the fragile equilibrium required for a sustainable water supply and the good functioning of groundwater-dependent ecosystems (GDE) [5–7]. In many densely populated and irrigated drylands, external water transfer from other basins [8,9], and inner production of non-conventional resources such as urban wastewater reuse and seawater and brackish groundwater desalination to cope with water scarcity are being encouraged [10–12].

When water is used in urban areas for supply and in peri-urban areas for irrigation agriculture, water fluxes from these areas to regional groundwater bodies occur. The mechanisms for urban wastewater leaching are known. Buried urban sanitation infrastructure might not be watertight due to deterioration, and a fraction of wastewater may leach before reaching the urban wastewater treatment facility. Irrigated agriculture also entails an unavoidable loss of water during irrigation that may reach the aquifers. On favorable topographic and hydrogeological conditions, for instance when steep topography and low-permeability bedrock exist, urban and irrigation returns can be identified in water quantity and quality changes of local springs. Although these mechanisms are well known, the groundwater resource in most medium-sized urban and peri-urban areas is unknown because they are typically ungauged settings. Therefore, conceptualization of the hydrogeological functioning of these areas is crucial to understand and evaluate how these water fluxes can be used to alleviate the pressure on many stressed coastal groundwater bodies. However, geological exposure in urban areas is typically low to define aquifer geometry, especially in flat areas where geological exploration and groundwater monitoring is restricted to some geotechnical soundings and pumping wells at most. The opposite happens in towns with steep topography where geological formations and groundwater dynamics are partially observable and evaluations are possible. These evaluations may serve to conceptualize the groundwater resource generated in neighboring towns having similar habits for water consumption, but a flat topography that prevents direct groundwater observations.

Both steep topography and hydrogeological exposure are found in Adra town in southern Spain. Adra town is placed in a steep versant over small aquifers that contribute to the Adra River Delta Groundwater Body (ARDGB), which sustains GDEs, and includes a northern peri-urban area devoted to irrigation agriculture in greenhouses [13–15].

On the basis of existing or compiled datasets, the groundwater resource evaluation involves three general stages [16–18]: (1) Aquifer geometry and hydraulics definition, (2) conceptualization of the hydrogeological functioning, and (3) evaluation of the water balance components. This paper is aimed at developing the above two first general stages in deep, whilst the third one evaluates the groundwater resource contributed from the northern steep urban sector (NSUS) to the southern flat urban sector (SFUS), which belongs to the ARDGB.

For aquifer geometry definition, previous geological information [19] was revisited and near-surface geophysical techniques were applied. Near-surface geophysical techniques have been widely used in groundwater research to acquire basic information on aquifer geometry and some transient hydrogeological features [20–25]. This paper combines the Multichannel Analysis of Surface Waves (MASW) [26–32] and Ground Penetrating Radar (GPR) [33] techniques for these purposes. A basic Darcy's Law formulation [34,35] was used to assess the NSUS groundwater discharge. For a fluent reading, the description for acronyms used is in Table 1.

Table 1. Description for acronyms used.

Acronym	Description
AGS	Actual groundwater storage
ARB	Adra River basin
ARD	Adra River Delta
ARDGB	Adra River Delta Groundwater Body

Table 1. *Cont.*

Acronym	Description
g1	Pleistocene colluvial 1
g2	Pleistocene colluvial 2
g3	Pleistocene colluvial 3
GDE	Groundwater-dependent ecosystem
GPR	Ground Penetrating Radar
MASW	Multichannel Analysis of Surface Waves
NSUS	Northern steep urban sector
PGS	Potential groundwater storage
SFUS	Southern flat urban sector
t1	Pleistocene littoral facies 1
t2	Pleistocene littoral facies 2
VE	Velocity of electromagnetic waves
VS	Shear-wave velocity
WB	Paleozoic weathered, fissured crystalline bedrock

2. Study Area

2.1. Location and Climate

Adra town (36°44′30″–36°45′30″ N, 3°00′–3°02′ W) is located at the outlet of the Adra River basin (ARB) in the province of Almeria in southern Spain (Figure 1a). This coastal town has a surface of 1.82 km^2 and is crossed by small temporary streams that flow to the sea (Figure 1b). From a geomorphological point of view, these urban basins delimit a peri-urban area (2.64 km^2 in the 50–436 m a.s.l. elevation range) that influences the hydrology of a fraction of the urban area (0.97 km^2 in the 0–50 m a.s.l. elevation range) downstream [15]. The peri-urban area is divided into a southern sector (1.41 km^2 in the 50–150 m a.s.l. elevation range) devoted to irrigation agriculture in greenhouses and a northern sector (1.23 km^2 in the 150–436 m a.s.l. range) with low human influence (Figure 1c).

Climate is warm-summer Mediterranean according to the Köppen classification [36], which means a semiarid regime with hot dry summers and temperate rainy winters [14]. Insolation is high, 2900 h per year in low-lying places.

The 0.01° (~1-km) resolution nodal daily precipitation and temperature (maximum and minimum) series from the Iberia01 grid over the period 1971–2015 [37] was used to deduce weather conditions. Precipitation (P) occurs in three distinctive phases, each about four-month long as (1) a predominant rainy phase from October to January which represents around 50% of annual P; (2) a moderately rainy phase from February to May which means around 40% of annual P; and (3) a dry phase from June to September which records around 10% of annual P. Extreme rainfall events over 80 mm per day have been documented. Temperature (T) shows a bimodal distribution, each period about six-month long as (1) a warm period from May to October with average minimum and maximum monthly T of 15 °C in October and 31 °C in July, respectively; and (2) a temperate period from September to April with average minimum and maximum monthly T of 9 °C in January and 20 °C in April, respectively.

2.2. Geological and Hydrogeological Setting

The study area (Figure 1b) belongs to the Alpujárride tectonic complex from the Internal Domain of the Alpine Betic Cordillera [38,39]. The area is tectonically active as a consequence of the convergence between the African and Eurasian Plates, which ended with the collision of the Internal and External Betic domains during the early Miocene [39,40]. The combination of active tectonics and sea-level changes controls the accommodation space for Upper Miocene to present sedimentation [39,41].

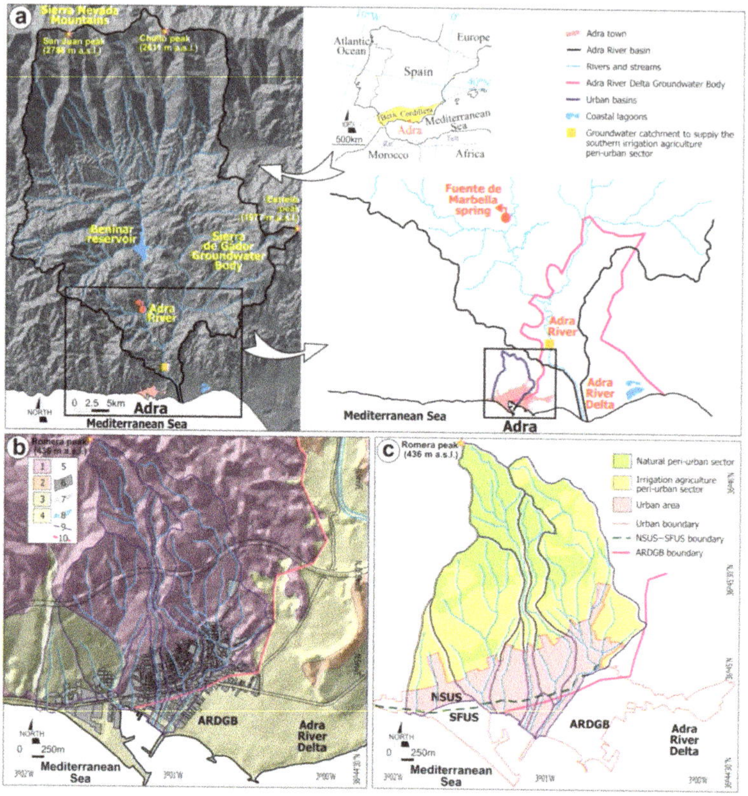

Figure 1. (**a**) Location of the study area in southern Spain, showing geographical sites cited in the text. (**b**) After the Geological Survey of Spain [38], the simplified 1:50,000-scale geological map of the urban and peri-urban areas; legend as (1) Paleozoic metamorphic formations, (2) Pliocene formations, (3) Pleistocene formations, (4) Holocene formations, (5) Undifferentiated geological contact, (6) Urban area, (7) Roads, (8) Adra River, (9) Urban basins, and (10) ARDGB boundary. (**c**) After the Andalusian Environmental Information Network [http://www.juntadeandalucia.es/medioambiente/site/rediam], the 1:25,000 scale land-use map of the urban basins, accessed on 15 January 2021.

Alcalá et al. (2002) [19] classified the geological record of Adra town into twelve geological formations attending to age, origin, and geometry. The two pre-orogenic formations included Paleozoic low- and medium-grade mica-schists and quartz-schists (crystalline bedrock). The ten post-orogenic formations included Pliocene to Quaternary sedimentary formations unconformably deposited over the bedrock (Figure 1b) as (1) a Pliocene deltaic formation; (2) five Pleistocene formations including two generations of littoral facies (t1 and t2) and colluvial (g1 and g2), and a cemented colluvial (g3); and (3) four Holocene formations including colluvial, the Adra River Delta (ARD) alluvial, present littoral facies, and anthropogenic filling. Direct field observations proved that the official 1:50,000-scale geological mapping [38] is detailed enough to explain the hydrogeological functioning in the peri-urban area (Figure 1b).

From a regional hydrogeological point of view, the eastern and southern sectors of Adra town are emplaced on the ARDGB, which is located at the ARB outlet (Figure 1a). The ARDGB has a surface of 49.2 km^2 and an average saturated thickness of 100 m. The average groundwater recharge and discharge are around 25 Mm^3 $year^{-1}$ [14,42–44].

In the NSUS, Paleozoic weathered, fissured crystalline bedrock (WB), Pleistocene t1 and t2, and Holocene colluvial form a marginal aquifer not officially catalogued as a groundwater body despite its historical relevance for water supply at homes and sparse traditional irrigated crops [15]. These formations are the subject of this paper.

2.3. Land Use and Water Allocation

In the area covered by the urban basins (Figure 1c), 1.29 km^2 is occupied by marginal rainfed crops, scrublands, and bare bedrock in the northern peri-urban sector, 1.41 km^2 is devoted to irrigation agriculture in greenhouses in the southern peri-urban sector, and 0.97 km^2 is urban area (Figure 1c). Irrigation agriculture in greenhouses is the main economic driver [13], which has attracted a new population in recent years. As a consequence, new urbanizations have occupied the fertile plain of the ARD in the eastern and southern sectors of the town (Figure 2). The Spanish National Institute for Statistics [45] reported around 20,000 inhabitants in the main urban area in 2019.

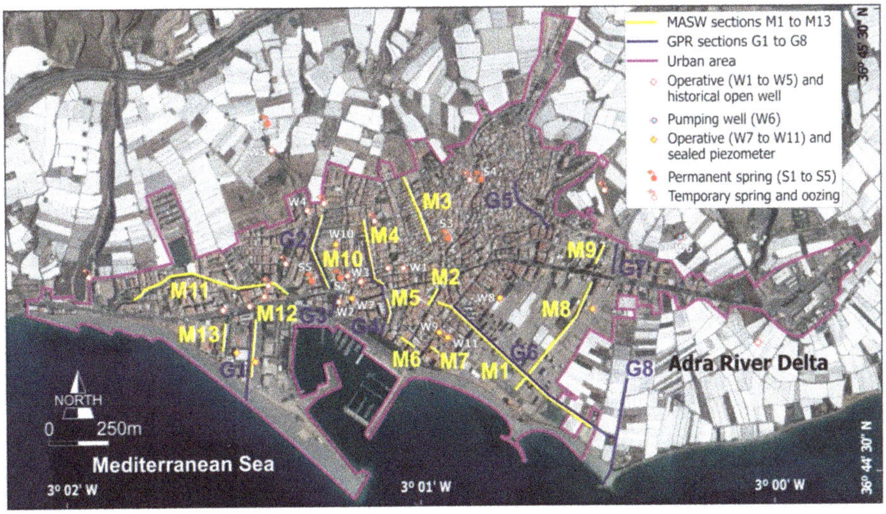

Figure 2. Satellite image of Adra town showing location of the MASW and GPR profiles, and selected groundwater observation points.

Water use is mostly devoted to urban and irrigation agriculture supply. Water allocation for urban supply is around 1.24 Mm^3 per year; 0.97 Mm^3 is intended to water supply at homes, 0.16 Mm^3 to auxiliary industry and services, and 0.11 Mm^3 to watering public gardens and urban cleaning. For 20,000 inhabitants, the average urban water endowment is 170 L per inhabitant and day [46]. Urban return from leakage in the sanitation network is around 0.20, after personal communication from the local water authority. Urban water allotment combines the ARDGW and the Fuente de Marbella spring from the Sierra de Gádor Groundwater Body in the mid-valley ARB (Figure 1a) [15]. In the peri-urban area, average water allocation for irrigation agriculture is around 7850 m^3 per hectare and year, from which 1800 m^3 are devoted to soil disinfestation in June when intensive crops rotate. Greenhouses produce vegetables (pepper, eggplant, cucumber, and similar others) in the winter–spring season, and melon and watermelon in the autumn–winter season. Irrigation is done through drip systems, which reduce evaporation and infiltration losses. Average irrigation efficiency is 0.85, thus the average irrigation return is 0.15 [47]. Irrigation water comes from groundwater pumped in the lower-valley Adra River alluviums, which belongs to the ARDWB (Figure 1a). Urban supply and irrigation agriculture do not use groundwater from local aquifers in the NSUS, although some abandoned handmade open wells evidence the historical use of this resource [15].

3. Methods

3.1. Aquifer Geometry Definition

3.1.1. Hydrogeological Field Surveys

Over the base of previous geological [19], hydrological [48], and hydrogeological [15] findings, and attending to the permeability type and storability reported by official reports [43,44,46], technical reports [49,50], and the scientific literature [14,42,51,52], the hydrogeological behavior of the geological formations is defined. Near-surface geophysical surveys (MASW and GPR) and new hydraulic tests to refine the geometry and hydrogeological functioning of the NSUS were performed. Two flash campaigns for piezometry and flow measurements in selected open wells (W1 to W6; Figure 2) and springs (S1 to S5; Figure 2) were carried out in September 2014 and June 2015. Piezometry used a level probe from Seba Hydrometrie with a precision of 0.005 m, whereas springs flow was measured manually 10-times to provide a confident average value.

3.1.2. MASW Technique

MASW is a seismic geophysical technique in which the Rayleigh wave fundamental mode dispersion curve and higher modes (if present) are extracted from a shot record and then inverted to generate a 1D VS [L T^{-1}] model [31,32]. This technique assesses the fundamental and higher modes simultaneously, thus permitting to obtain more accurate VS models [29,30]. A roll-along setup with a land-streamer acquisition system was used for data acquisition.

MASW data were acquired using a 24-channel SUMMIT II Compact Seismograph by DMT, Germany, with the following configuration: Recording array of 24 vertical component geophones, 2-m geophone spacing, 4-m separation between the source impact point and first geophone to minimize near-source effects, 2 stacks, 10-m displacement between readings, and a sampling rate of 0.25 ms. A Wacker Neuson BS60-4s vibratory rammer was used to generate the Rayleigh waves.

Data analysis was carried out with SurfSeis3 software® by the Kansas Geological Survey, The University of Kansas, USA. Data processing consisted of geometry edition, data filtering, muting (when needed), generation of overtones (frequency–time energy diagrams), and fundamental and higher modes (if present) identification. Finally, dispersion curves were determined and then subjected to a mathematical inversion process to obtain a continuous 2D VS model. Additional methodological details can be consulted in Martínez-Pagán et al. (2018) [53].

3.1.3. GPR Technique

GPR is an electromagnetic geophysical technique which uses transmitting and receiving antennas to record the time of propagation of the electromagnetic signal in the subsoil. In this study, the common or single offset antenna setup was used [33]. This technique provides radargrams, which are a set of radar traces, each representing the acquisition of the reflected signal in a point of the ground surface in time [54]. The propagation velocity of electromagnetic waves (VE) [L T^{-1}] and their amplitudes through the subsurface depends on the electrical and magnetic properties of geological materials and the adopted antenna frequency [55–58]. Penetration depth of the electromagnetic signal decreases as the clay content and salinity of the media and the antenna center frequency increase [33,55,58–61].

A GSSI SIR-3000 system equipment with a 270-MHz shielded antenna mounted on a cart with an encoder was used for data acquisition. The Reflexw software by Sandmeier was used for data processing. Relative dielectric permittivity was set to 10 according to near-surface subsoil characteristics, and later adjusted during processing. Processing flow consisted of time-zero correction, velocity refinement through comparison with well-known site features such as water table and bedrock depths, background removal, 1D filtering—bandpass butterworth filter maintaining the 70–400 MHz range, and topography handling. Additional methodological details can be consulted in Paz et al. (2007) [33].

3.2. Groundwater Resource Evalution

In the NSUS, groundwater exploitation is virtually null, and groundwater discharge can be used as a reliable proxy of the groundwater resource contributed downstream. A basic Darcy's Law formulation [34,35] for groundwater discharge was implemented, as:

$$D = -K\, dh/dl = i\, K\, b\, l, \quad (1)$$

where i is dimensionless hydraulic gradient; K is permeability expressed as the water flow traveled distance per time [L T^{-1}], in this case m day^{-1}; b is aquifer saturated thickness [L] in m; and l is aquifer discharge section [L] in m.

4. Results

4.1. Geophysical Data

4.1.1. MASW Survey

In February 2014, Martínez-Pagán et al. (2018) [53] completed thirteen MASW sections (labelled from M1 to M13, Figure 2). The MASW survey covered all the geological formations defined in Adra town. Basic data for all VS models were frequency in the 2.8–43.0 Hz range, phase velocity in the 259–1198 m s^{-1} range, length from 40 m (section M7) to 810 m (section M1), and prospecting depth from 43 m (section M9) to 84 m (section M1). From active and passive MASW measurements, the 1D VS models were generated and interpolated to create 2D VS sections (called MASW sections) (Figure 3). The sections were topographically corrected. The vertical-equispaced VS values from all MASW sections were georeferenced and interpolated to create 2D VS layers at different elevations (called MASW maps) (Figure 4).

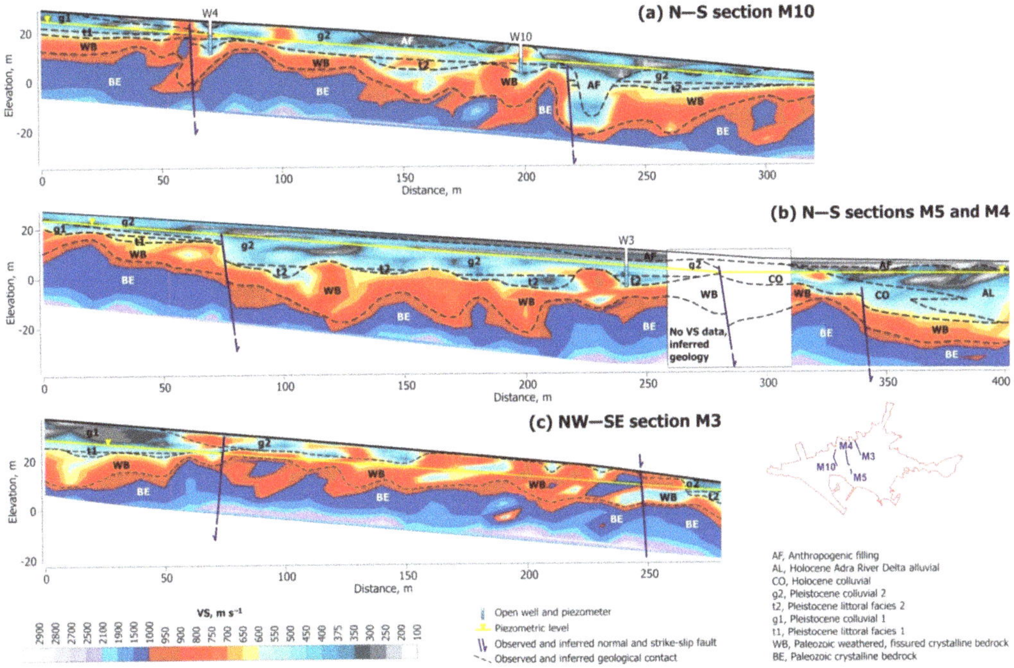

Figure 3. MASW sections M10, M4, M5, and M3; location is in Figure 2. A geological interpretation of VS values after Alcalá et al. (2002) [19] and the projected piezometric level after the flash campaign carried out in September 2014 are included; location of open well and piezometer W3, W4, and W10 is in Figure 2. Sections are topographically corrected and its vertical-to-horizontal scale ratio is 1:1.

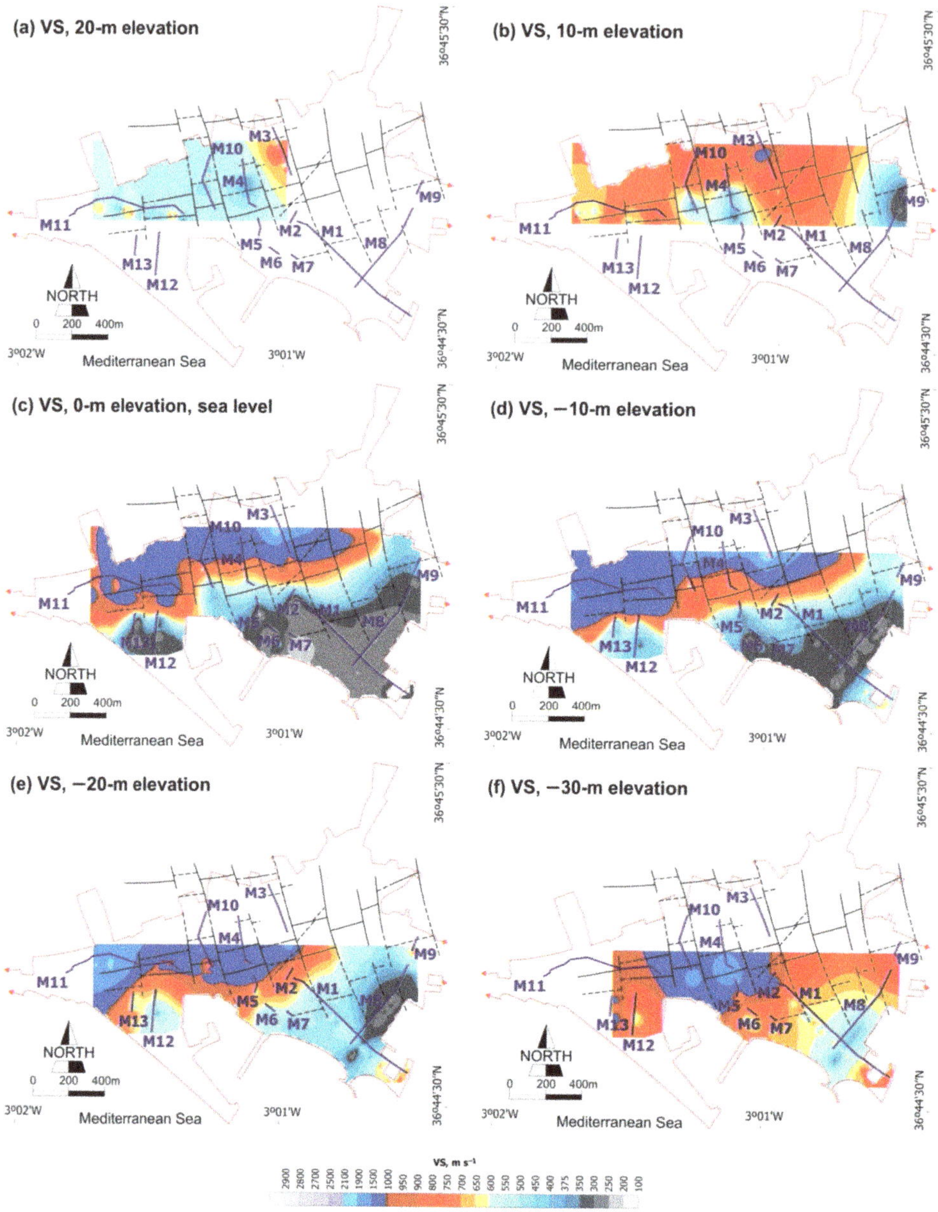

Figure 4. Some MASW maps at different elevation regarding the sea level in Adra town, showing observed and inferred normal and strike-slip fault systems, and location of MASW sections M1 to M13.

This paper uses the MASW sections and maps for geological definition following the interpretative criteria reported by Paz et al. (2020) [62]. These authors propose that VS propagation in sediments is a site-specific steady property determined by effective compaction and as such it is dependent on the age and depth of each geological material piled on vertical [63–66]. In Adra, the VS values obtained in previous studies [19,53,67] were <350 m s^{-1} for Holocene sediments, 350–600 m s^{-1} for Pleistocene sediments, 600–900 m s^{-1} for

Pliocene sediments, 900–1200 m s^{-1} for WB, and >1200 m s^{-1} for bedrock. Some sedimentary and tectonic processes can modify these ranges, such as cementation of coarse-grained sediments increasing VS up to 800 m s^{-1} or high fissuring and weathering of the bedrock decreasing VS up to 700 m s^{-1} [19,67]. For hydrogeological interpretation, the available geological and hydrogeological information was superimposed on sections M1 to M13. Four sections (M10, M4, M5, and M3) of special interest to deduce the hydrogeological functioning of the NSUS are displayed in Figure 3.

Sections M10 (Figure 3a), M4 (Figure 3b), and M3 (Figure 3c) show the NSUS, whereas section M5 (Figure 3b) shows the SFUS. Due to their continuity, sections M4 and M5 were interpreted together into a single section. A succinct description of these MASW sections is below. The uppermost 1–8 m thick VS < 300 m s^{-1} is attributed to anthropogenic filling (sections M10, M4, M5, and M3), the underlying 2–10 m thick VS > 350 m s^{-1} to Holocene colluvial and present littoral facies (section M5), the underlying 10–15 m thick 300 < VS < 600 m s^{-1} to clay-rich g1 and g2 and coarse grained t1 and t2 (t1 is not identified in section M3), the underlying 5–10 m thick 600 < VS < 900 m s^{-1} to cemented levels at the bottom of t1 and t2, the underlying 15–20 m thick 900 < VS < 1200 m s^{-1} to WB, and the deeper VS > 1200 m s^{-1} to the bedrock. This overall vertical VS distribution correlates well with the geological information, but some disruptions are observed. In M10, punctual VS < 300 m s^{-1} underlying t2 is due to a dig gallery built in the 19th century to drain groundwater. In M10, M4, and M3, some patches embedded into g1 and g2 with VS > 1000 m s^{-1} are attributed to isolated bedrock blocks. The pairs g1–t1 and g2–t2 show similar 300 < VS < 600 m s^{-1}, thus limiting identifications of the upper boundary of t1 and t2 forming aquifers regarding the lower boundary of g1 and g2 forming aquitards. The lowermost part of t1 and t2 and the uppermost part of WB are cemented and show similar VS < 900 m s^{-1}, thus preventing its boundary definition. As described in next Section 4.1.2, radargrams help to disambiguate these boundaries.

MASW maps (Figure 4) show how spatial VS continuity is interrupted by NW–SE normal and NE–SW strike-slip fault systems associated to active tectonics. The former is a first-order fault system determining the accommodation space for the Quaternary sedimentation whereas the latter is a second-order one that compartmentalizes the bedrock, thus inducing interruptions of some geological formations such as t1 and t2. The MASW maps identify geological boundaries of special hydrogeological interest, such as the (1) bedrock bathymetry at the 0-m (Figure 4c), −10-m (Figure 4d), and −20-m (Figure 4e) elevation maps; (2) NSUS–SFUS boundary at the 0-m (Figure 4c) and −10-m (Figure 4d) elevation maps; and (3) WB extension at the 10-m elevation map (Figure 4b).

4.1.2. GPR Survey

In June 2015, Paz et al. (2017) [33] completed eight GPR sections (labelled from G1 to G8, Figure 2). The GPR survey covered all the geological formations and coarsely followed the trace of the MASW sections to (1) disambiguate the boundary of geological structures having similar VS, such as the pairs t1–g1 and t2–g2; (2) deduce thickness of the shallowest anthropogenic filling; and (3) delineate transient hydrogeological features such as water table, capillary fringe, and seawater–freshwater interface. Basic data for all GPR sections were 50-Hz T-rate, 150-ns range, 120 scans per second, 40 scans per meter, 512 samples per scan, length from 88 m (section G3) to 652 m (section G6), and 4.5-m prospecting depth. Radargrams were topographically corrected and hydrogeological data were superimposed to refine shallow hydraulic features and geometry of the geological formations.

Three GPR sections (G2, G3, and G4) of special interest to deduce the hydrogeological functioning of the NSUS are displayed in Figure 5. In Adra, steep topography determines sub-horizontal reflections for Pleistocene sedimentation in the NSUS and horizontal ones for Holocene sedimentation in the SFUS. The southern sector of section G2 (Figure 5b) and the central sector of section G4 (Figure 5c) show how t2 produces stronger reflections than g2 due to cementation, thus permitting its identification. Outside the displayed sectors, the pair t1–g1 shows the same behavior. Bedrock and WB produce similar strong reflections, thus limiting its boundary definition in some cases. Some shallow hydrogeological features are identified. An easily detected reflector inside the clay-rich g1 and g2 in G2 (Figure 5b) and G4 (Figure 5c) is capillary fringe, which narrowly follows piezometry measured in wells W1, W3, and W10 (Figure 2). This means that t1, t2, and the uppermost part of the WB together form an unconfined aquifer whose upward groundwater flow determines the capillary fringe position inside the clay-rich g1 and g2, which form an aquitard.

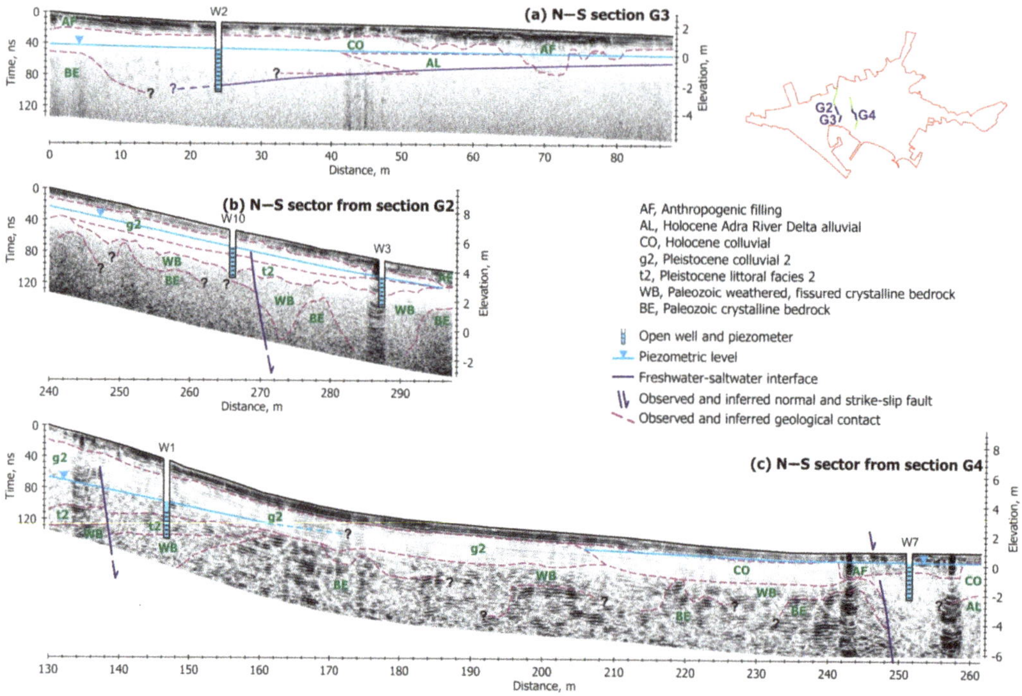

Figure 5. Selected sectors from GPR sections G2, G3, and G4 of especial hydrogeological interest; location is in Figure 2. The interpretation of radargrams was based on previous geological [19] and hydrogeological [15] data, and MASW sections (Figure 3). The projected piezometric level after the flash campaign carried out in June 2015 is included; location of open well and piezometer W1, W2, W3, W7, and W10 is in Figure 2. Profiles are topographically corrected and its vertical-to-horizontal scale ratio is 1:2.

In the SFUS, section G3 (Figure 5a) and the southern sector of section G4 (Figure 5c) show how the Holocene ARD Formation produces strong reflectors. In these coarse-grained sediments, the first strong reflection associated to the saturated media detection is stronger than the observed one in g1 and g2 and very closer to the water table, as deduced from piezometry measured in wells W2 and W7 (Figure 2), thus corroborating the unconfined behavior of this hydrogeological formation. In section G3 (Figure 5a), the VE signal loss below the sea level is attributed to the seawater–freshwater interface.

4.2. Hydrogeological Conceptualization

This section completes the two first goals of this paper, i.e., geometry and hydraulics definition of the geological formations in the NSUS. For geometry definition, geophysical findings, and previous [19] and new geological data were integrated on GIS to prepare the 1:5000 scale hydrogeological map of Adra town (Figure 6) and three representative hydrogeological cross-sections (Figure 7). For hydraulics definition, a permeability and effective porosity database was prepared from (1) compiled data from the literature described in Section 3.1.1, which included official publications devoted to the Holocene ARD alluvial Formation in the SFUS, technical reports surveying the NSUS and SFUS, and scientific publications covering the NSUS and SFUS; and (2) data from specific field surveys in the NSUS, which included five double-ring infiltration tests in low-permeability formations, three pumping tests in open wells in high-permeability formations, and nine granulometric curves in different formations. On the basis of this information, the twelve geological formations have been classified into ten hydrogeological formations attending to the permeability type and storability (Table 2). The potential groundwater storage (PGS) of each hydrogeological formation is defined as the product of surface (direct and underlying outcrops), thickness (saturated and unsaturated), and effective porosity (Table 2). PGS must be considered a tentative magnitude of the maximum storability, not the actual storability. PGS has been calculated for all formations in the total urban area and for the existing ones in the NSUS.

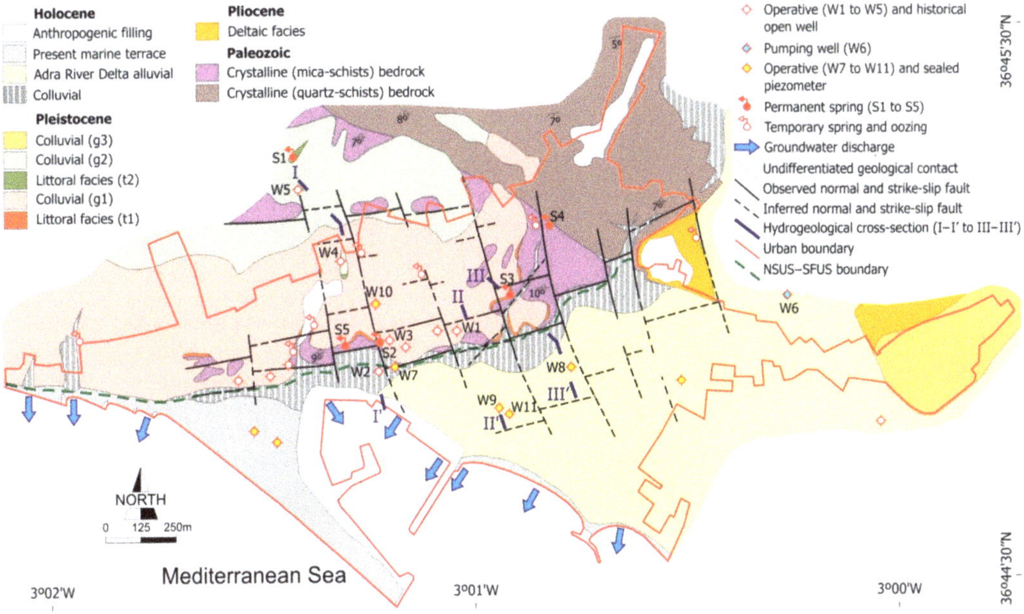

Figure 6. Hydrogeological map of Adra town at scale 1:5000, showing operative and historical groundwater observation points, sites where groundwater discharge to the sea is observable, and hydrogeological cross-sections I–I', II–II', and III–III' as in Figure 7.

The Paleozoic crystalline (mica-schists and quartz-schists) bedrock is a low-permeability formation constituting the impervious base of aquifers in the urban area (Table 2). This is catalogued as aquiclude. The MASW sections (Figure 3) and maps (Figure 4) delineate the bedrock geometry through VS > 1200 m s^{-1}.

The WB Formation is a porous media forming a moderate- to high-permeability aquifer. In the NSUS, the surface is 0.68 km^2, thickness is in the 1–18 m range, and its average PGS is around 0.27 Mm^3 (Table 2). The geometry of WB can be deduced through VS in the 800–1200 m s^{-1} range from MASW sections (Figure 3) and maps (Figure 4). Recharge comes from direct rainfall and runoff infiltration, and urban and irrigation returns.

The Pliocene deltaic facies formation is unconformably deposited over the bedrock in the SFUS. This formation is catalogued as a moderate- to high-permeability aquifer (Figure 6) of 0.09 km^2, thickness in the 4–31 m range, and average PGS around 0.03 Mm^3 (Table 2). Recharge comes from transference from the Holocene ARD alluvial formation.

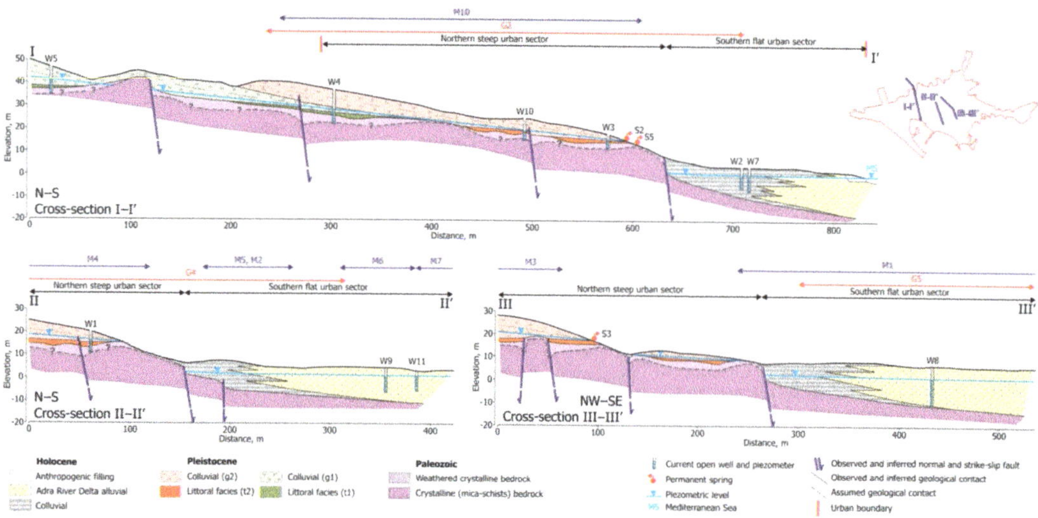

Figure 7. Three hydrogeological cross-sections I–I', II–II', and III–III' (see location in Figure 6), showing the distances covered by MASW (double sided arrow blue lines) and GPR (double sided arrow red lines) sections, operative and historical groundwater observation points, and the projected piezometric level after the flash campaign carried out in June 2015. Cross sections are topographically corrected, and its vertical-to-horizontal scale ratio is 1:2.

The Pleistocene record includes two littoral–continental sedimentary sequences in the NSUS and a continental sequence in the SFUS (Figure 6). A brief description is below.

In the NSUS, the two littoral–continental sequences are unconformably deposited over the bedrock (Figure 7). Each sequence includes littoral facies (t1 and t2) underlying clay-rich colluvial (g1 and g2) (Figures 6 and 7). t1 and t2 form a high-permeability aquifer of 0.41 km^2, thickness in the 1–8 m range, and average PGS around 0.21 Mm^3 (Table 2). t1 and t2 are hydraulically connected to WB and partially disconnected between them, as deduced from the studied MASW (Figure 3) and GPR (Figure 5) sections. The normal and strike-slip fault systems compartmentalize t1 and t2, but do not interrupt apparently the continuity of WB, as deduced from the 10-m elevation MASW map (Figure 4b). The result is groundwater flowing throughout WB, t1, and t2, thus forming together a confined aquifer in the NSUS. Geometry, piezometry in open wells, and groundwater discharge in springs at different elevations corroborate this confined hydraulic behavior (Figure 7). This aquifer is hydraulically connected (and discharges) to the unconfined Holocene ARD alluvial formation. Recharge comes from direct rainfall and runoff infiltration, and urban and irrigation returns. g1 and g2 are low-permeability formations catalogued as aquitards that confine t1 and t2, respectively (Figures 6 and 7). Surface is 0.58 km^2, thickness is in the 1–32 m range, and its average PGS is around 0.04 Mm^3 (Table 2).

In the SFUS, the third Pleistocene continental sequence includes a cemented colluvial (g3) unconformably deposited over the Pliocene deltaic facies Formation. This is a moderate- to low-permeability formation cataloged as an aquitard (Figure 6). Surface is 0.10 km^2, thickness is in the 1–29 m range, and its average PGS is around 0.03 Mm^3 (Table 2). Recharge comes from direct rainfall and runoff infiltration, and transference from the Holocene ARD alluvial formation.

The Holocene sedimentary record includes four formations: Colluvial, ARD alluvial, present littoral facies, and anthropogenic filling. A brief description is below.

The colluvial formation is emplaced on the NSUS–SFUS boundary unconformably deposited over the bedrock and the pair t2–g2 (Figures 6 and 7). It is catalogued as a moderate- to high-permeability aquifer. In the NSUS, surface is 0.11 km^2, thickness is in the 1–21 m range, and its average PGS is around 0.07 Mm^3 (Table 2). Recharge comes from discharge from upstream hydrogeological formations, direct rainfall and runoff infiltration, and urban return. Discharge is done to the Holocene ARD alluvial formation and the sea.

The ARD alluvial Formation is in the SFUS and belongs to the ARDGB. It is cataloged as a high-permeability aquifer (Figure 6). The surface is 0.59 km^2, thickness is in the 1–35 m range, and its average PGS is around 0.78 Mm^3 (Table 2). Recharge comes from upstream infiltration in the Adra River valley alluviums, and irrigation and urban returns. Discharge is produced by pumping, transference to other hydrogeological formations, and to the sea.

The present littoral facies formation is in the SFUS and results from civil works to prevent coastal erosion [19]. Only the western beaches have hydrological interest, forming a shallow high-permeability aquifer of 0.17 km^2, thickness in the 1–9 m range, and average PGS of around 0.09 Mm^3 (Table 2). Recharge comes from rainfall and runoff infiltration, and discharge from other upstream hydrogeological formations.

The anthropogenic filling formation appears as high-permeability patches that contribute to urban runoff or groundwater storage when overlying low- or high-permeability formations, respectively. In the NSUS, surface is 0.05 km^2, thickness is in the 1–6 m range, and the average PGS is around 0.03 Mm^3 (Table 2). Recharge comes from rainfall and runoff infiltration, and urban return.

4.3. Groundwater Resource Evaluation

This section completes the third goal of this paper, i.e., the groundwater resource evaluation in the NSUS. Excluding formations behaving as aquicludes (bedrock) and aquitards (g1, g2, and g3), the average PGS in Adra town is around 1.77 Mm^3. This figure varies in the 0.12–5.49 Mm^3 range when the minimum and maximum thickness and effective porosity values are used (Table 2). In the NSUS, g1 and g2 confine the aquifer formed by WB, t1, and t2. Excluding bedrock, g1, and g2, average PGS is around 0.58 Mm^3 and its range is 0.05–1.71 Mm^3 (Table 2). The actual groundwater contribution to the SFUS must be less than average PGS, which is the theoretical maximum value.

Since groundwater exploitation is virtually null, the NSUS groundwater discharge can be considered a reliable proxy of the groundwater contributed from the NSUS to the SFUS. For groundwater discharge evaluation through the Darcy's Law formulation, the hydrogeological formations in the NSUS were grouped into three aquifers (1) that formed by WB, t1, and t2 (called NSUS aquifer); (2) the northern sector of the colluvial formation; and (3) the anthropogenic filling patches. In each aquifer, input data were saturated thickness, discharge section, hydraulic gradient, and permeability (Table 3).

Average saturated thickness is total thickness minus the difference in topography and piezometry deduced from open wells and springs. Average saturated thickness of the NSUS aquifer is around 0.85-fold the average total thickness described in Table 2. This figure results from weighting full-saturated aquifer sectors such as those shown in the hydrogeological cross-sections I–I′ to III–III′ (Figure 7) and others fully desaturated. Average saturated thickness of colluvial and anthropogenic filling formations are 0.7- and 0.5-fold the average total thickness, respectively. The aquifer discharge sections were deduced from the hydrogeological map (Figure 6). The desaturated aquifer sectors were excluded from this calculation.

After the two flash campaigns carried out in September 2014 and June 2015, piezometry in open wells W1 to W6 (Figure 2) and groundwater discharge elevation in springs S1 to S5 (Figure 2) were used to delineate the piezometric level and define the hydraulic gradient. Average hydraulic gradients varied in the 0.017–0.044 range and coarsely followed the topographic gradient as 0.035 in the NSUS aquifer, 0.017 in the colluvial formation, and 0.004 in the anthropogenic filling formation (Table 3). Average permeability for these formations is in Table 2.

Table 2. Geometry, hydraulics data, and potential groundwater storage (PGS) of the hydrogeological formations in the total urban area and the NSUS.

Age	Hydrogeological Formation	Main Lithology	Urban Surface, km [1,2]	NSUS Surface, km [1,2]	Thickness, m [2]	Permeability, m day^{-1} [3]	Effective Porosity, [3]	PGS in the Urban Area, Mm [3,4]	PGS in the NSUS, Mm [3,4]	Hydraulic Behavior
Paleozoic	Bedrock	Mica-schists, quartz-schists	1.82	0.77	–	–	–	–	–	Aquiclude
	Weathered, fissured bedrock	Mica-schists, quartz-schists	0.86	0.68	1–18 (8)	0.9–1.5 (1.1) n = 3	0.02–0.07 (0.05) n = 3	0.02–1.08 (0.34)	0.01–0.86 (0.27)	Aquifer
Pliocene	Deltaic facies	Sand, gravel, silt	0.09	–	4–31 (17)	1.1–5.2 (2.9) n = 3	0.01–0.05 (0.02) n = 3	<0.01–0.14 (0.03)	–	Aquifer
Pleistocene	Littoral facies (t1, t2)	Gravel, sand	0.41	0.41	1–8 (4)	2.3–9.3 (4.2) n = 4	0.06–0.17 (0.13) n = 4	0.03–0.56 (0.21)	0.03–0.53 (0.21)	Aquifer
	Detrital colluvial (g1, g2)	Gravel, sand, clay	0.58	0.58	1–32 (9)	0.05–0.09 (0.07) n = 3	<0.01–0.01 (<0.01) n = 3	<0.01–0.20 (0.04)	<0.01–0.20 (0.04)	Aquitard
	Cemented colluvial (g3)	Cemented gravel, sand	0.10	–	1–29 (14)	0.12–0.44 (0.28) n = 2	0.01–0.03 (0.02) n = 2	<0.01–0.09 (0.03)	–	Aquitard
Holocene	Adra River Delta alluvial	Sand, silt	0.35	0.11	1–21 (9)	0.8–2.9 (1.4) n = 3	0.05–0.10 (0.07) n = 3	0.02–0.74 (0.22)	<0.01–0.23 (0.07)	Aquifer
	Present littoral facies	Sand	0.59	–	1–35 (19)	5.3–30.7 (13.8) n = 6	0.04–0.12 (0.07) n = 6	0.02–2.48 (0.78)	–	Aquifer
	Colluvial	Gravel, sand, clay	0.17	–	1–9 (4)	5.1–30.3 (14.5) n = 4	0.08–0.17 (0.13) n = 4	0.01–0.26 (0.09)	–	Aquifer
	Anthropogenic filling	Blocks, sand, silt	0.17	0.05	1–6 (3)	5.2–8.2 (6.7) n = 2	0.14–0.23 (0.17) n = 2	0.02–0.24 (0.09)	<0.01–0.07 (0.03)	Aquifer

[1] Surface of Adra town is 1.82 km^2 and surface of the NSUS is 0.77 km^2. Surface of each hydrogeological formation is the sum of direct and underlying outcrops deduced from the 1:5000 scale hydrogeological map (Figure 6) and cross-sections (Figure 7), geotechnical sounding data, MASW sections (Figure 3) and maps (Figure 4), and GPR sections (Figure 5). [2] Thickness is deduced from direct field observations, geotechnical sounding data, and geophysical surveys; in parenthesis is the average value calculated on GIS. [3] Permeability and effective porosity data come from the compiled data from the literature described in Section 3.1.1 and the specific field surveys described in Section 4.2; in parenthesis is the average value; n is the number of data. [4] PGS is the product of surface, thickness (saturated and unsaturated), and effective porosity of each hydrogeological formation; in parenthesis is the average value.

Table 3. NSUS groundwater discharge after the Darcy's Law formulation.

Hydrogeological Formation [1]	i [2]	K [2]	b [2]	l [2]	D [2]	AGS [2]
Weathered, fissured bedrock	0.035	0.9–1.5 (1.1) n = 3	0.8–15.3 (6.8)	1860	0.02–0.55 (0.18)	0.01–0.73 (0.23)
Littoral facies (t1, t2)	0.035	2.3–9.3 (4.2) n = 4	0.8–6.8 (3.4)	450	0.01–0.36 (0.08)	0.02–0.47 (0.18)
Detrital colluvial (g1, g2)	0.035	0.05–0.09 (0.07) n = 3	0.5–16.0 (4.5)	1610	<0.01–0.03 (<0.01)	<0.01–0.10 (0.02)
Colluvial	0.017	0.8–2.9 (1.4) n = 3	0.6–12.6 (5.4)	480	<0.01–0.11 (0.02)	<0.01–0.14 (0.04)
Anthropogenic filling	0.004	5.2–8.2 (6.7) n = 2	0.5–3.0 (1.5)	110	<0.01	<0.01–0.03 (0.01)

[1] In Table 2, age, lithological description, and PGS of each hydrogeological formation. [2] i = average dimensionless hydraulic gradient; K = permeability in m day^{-1} (Table 2); b = aquifer saturated thickness in m; l = aquifer discharge section in m; D = NSUS groundwater discharge in Mm^3 year^{-1}; and AGS = actual groundwater storage in Mm^3 as the product of b, surface (Table 2), and effective porosity (Table 2). K, b, D, and AGS include range and average value into parenthesis.

Excluding g1 and g2 behaving as aquitards, the average NSUS groundwater discharge is around 0.28 Mm^3 year^{-1}. This figure varies in the 0.03–1.02 Mm^3 year^{-1} range when the minimum and maximum values of saturated thickness and permeability are considered (Table 3). Average actual groundwater storage (AGS), which is expressed as the product of surface (direct and underlying outcrops as in Table 2), saturated thickness, and effective porosity (Table 2), is around 0.47 Mm^3 (Table 3).

NSUS groundwater discharge was compared to groundwater recharge produced from precipitation in the peri-urban area (2.64 km^2) and NSUS (0.77 km^2), irrigation return in the southern peri-urban sector devoted to irrigation agriculture in greenhouses (1.41 km^2), and urban return in the NSUS (Table 4). Using tracer and physical techniques, Alcala et al. (2008) [15] tentatively evaluated the average recharge rate from precipitation in the peri-urban and urban areas as 10 mm year^{-1}. This figure is similar to that reported by Andreu et al. (2011) [68] in coastal areas in southern Almería province. The average recharge from precipitation is 0.03 Mm^3 year^{-1}. As described in Section 2.3, average water allocation for urban supply is 1.24 Mm^3 year^{-1} [46] After checking a similar inhabitant density along the main urban area, this figure was linearly approached as 0.52 Mm^3 year^{-1} for the 0.77-km^2 NSUS. Applying the urban return coefficient of around 0.20 informed by the local water authority, the urban return contributing to recharge is around 0.10 Mm^3 year^{-1}. Average water allocation for irrigation agriculture is around 7850 m^3 per hectare and year and the average irrigation return coefficient is 0.15 [47]. The irrigation return contributing to recharge is around 0.17 Mm^3 year^{-1} (Table 4). Average groundwater recharge is around 0.31 Mm^3 year^{-1} (Table 4). This figure is 0.03 Mm^3 year^{-1} higher than average groundwater discharge.

Table 4. Groundwater recharge in the NSUS.

Recharge Component	Area Covered [1]	Surface, km^2	Recharge, Mm^3 year^{-1}	Reference
Precipitation	PUA, NSUS	3.41	0.03	[15]
Urban return	NSUS	0.77	0.10	[46]
Irrigation return	SPUS	1.41	0.17	[47]

[1] PUA = peri-urban area; NSUS = northern steep urban sector; SPUS = southern peri-urban area devoted to irrigation agriculture in greenhouses.

5. Discussion

5.1. The MASW Technique for Geological Definition

The MASW technique has widely been used in seismic hazard research, especially in urban areas [29–32,53,69]. The experience in shallow groundwater research is incipient [26–28] and has mostly focused to disambiguate geological structures when other near-surface geophysical techniques fail to obtain a reliable interpretation [26,27,62]. This paper widens the experience in shallow groundwater research to define the geometry of geological structures once reference VS values are available. In Adra town, the confident use of the MASW technique had the VS values assigned to each geological formation in previous research [19,53,67], e.g., VS > 1200 m s^{-1} for Paleozoic bedrock, 800–1200 m s^{-1} for WB, 500–900 m s^{-1} for Pliocene deltaic facies, 350–600 m s^{-1} for Pleistocene formations, and <350 m s^{-1} for Holocene formations.

However, the MASW technique must be used together with other near-surface geophysical techniques relying on other subsurface properties other than VS in order to solve potential constraints imposed by the possible similar VS response of different geological structures. For instance, g1 and g2 show similar VS values than t1 and t2 (Figure 3), meaning a handicap at defining the geometry of the confined NSUS aquifer formed by WB, t1, and t2 (Figure 7). Adra town is a low-resistivity coastal area having high environmental salinity and clay-rich formations such as g1 and g2, so the geophysical electrical techniques do not seem to be suitable for disambiguating the pairs g1–t1 and g2–t2. The GPR technique was chosen to disambiguate these structures having similar VS values, as well as to identify key hydrogeological features, such as piezometric level, capillary fringe, and seawater–freshwater interface (Figure 5). MASW provides higher exploration depths than GPR, but less detailed resolution, whereas GPR is highly responsive to detailed subsurface electrical and magnetic changes related to natural and human-induced geological and hydrological heterogeneities [33,55,58–60].

This paper introduces a novelty in VS data post-treatment. Several vertical-equispaced MASW maps from –35-m to 30-m elevation regarding the sea level were created to identify continuity of the geological structures. This VS data post-treatment has enabled us to display where and why the boundary between the bedrock and the Pliocene to Quaternary sedimentary formations changes at different depths, for instance, abruptly due to normal and strike-slip fault systems or smoothly due to sedimentary processes (Figure 4). The bedrock bathymetry determines the accommodation space for sedimentary formations forming aquifers. These applications are of particular interest in urban areas where the capability for direct subsoil observations is typically quite limited. In Adra town, MASW sections (Figure 3) and maps (Figure 4) helped to create the 1:5000 scale hydrogeological map (Figure 6) and cross-sections I–I' to III–III' (Figure 7). This paper proposes the MASW technique for geological definition in urban shallow groundwater research, taking into account that other techniques to solve possible constraints imposed by the explored media features may be needed.

5.2. Use of the Groundwater Resource

The sources and mechanisms for groundwater recharge and discharge in urban areas are more numerous and complex than in natural environments, as documented many urban groundwater research [70–75]. Buildings and civil works combine with human-made drainage networks, sanitation systems, and paving to introduce new recharge and discharge components or modify the existing ones. In Adra town, this problematic increase because the study area must be extended to include a peri-urban area devoted to natural uses and irrigation agriculture that influences the urban hydrology downstream. In the NSUS, irrigation (0.17 Mm^3 $year^{-1}$) and urban (0.10 Mm^3 $year^{-1}$) returns are clearly higher than the recharge from precipitation (0.03 Mm^3 $year^{-1}$). The consequence is an uncatalogued groundwater resource contributing to the SFUS. The ARDGB, where the SFUS is emplaced, provides most of the usable water, sustains some GDEs catalogued in the Ramsar Convention list [14], and is officially protected to avoid new exploitations [43,44,46]. However, the NSUS and the southern irrigation agriculture peri-urban sector form a not officially catalogued marginal hydrogeological system that contributes to the SFUS and therefore to the ARDGB. This groundwater resource discharges to the sea under Adra town. This paper conceptualizes the functioning of this marginal hydrogeological system and provides a preliminary evaluation of the groundwater resource contributed to the SFUS. This resource could sustainably be exploited downstream.

Average groundwater discharge is around 0.28 Mm^3 $year^{-1}$, and may vary in the 0.03–1.02 Mm^3 $year^{-1}$ range when the minimum and maximum values of saturated thickness and permeability are considered (Table 3). Average aquifer recharge is around 0.31 Mm^3 $year^{-1}$. As expected in unexploited aquifers, the magnitude of average groundwater discharge and recharge in the NSUS is similar and lower than the calculated AGS of around 0.47 Mm^3 (Table 3). Three conclusions regarding the conceptualization of the NSUS aquifer functioning are gained: (1) The recharge-to-discharge absolute difference is 0.03 Mm^3 $year^{-1}$, so other sources and mechanisms for groundwater discharge probably occur and must be characterized, for instance occasional groundwater pumping to lower the piezometric level during building construction, groundwater up-take by phreatophytes and deep-rooted vegetation, and direct groundwater evaporation in sites having a quite shallow piezometry; (2) average saturated thickness is about 0.5–0.9-fold the total thickness of the hydrogeological formations, but there are desaturated sectors that must be characterized; and (3) renewability of the groundwater resource is high enough to sustain a durable small exploitation downstream, as deduced from an average groundwater turnover time less than one year, here tentatively expressed by means of the groundwater recharge (0.31 Mm^3 $year^{-1}$; Table 4) to AGS (0.47 Mm^3; Table 3) ratio [76,77]. This exploitation may guarantee the permanent water supply to watering public gardens and urban cleaning, currently around 0.11 Mm^3 per year and tending to increase in coming years as new urbanized areas are being planned. Other ecological uses could be also considered.

Other hydrogeological gaps to resolve in future research are (1) definition of the hydraulic behavior of fault as water-bearing or water-tight, (2) characterization of the hydraulic effect of civil works on groundwater flow, and (3) how climate change and subsequent land-use adaptations can affect this groundwater resource.

6. Conclusions

The favorable climate in the Mediterranean coastal area has potentiated an increasing urbanization and occupation of peri-urban areas for profitable irrigation agriculture. The new land uses and water demands have evidenced the controversy of having a scarce conventional water resource, while the intensive water use generates a difficult-to-manage non-conventional water resource that may complement the conventional one. In the context of global change and growing water demands, the small marginal aquifers in urban and peri-urban areas may play a role in complementing the urban allotment for specific uses. The smart cities in the near future will consider new paradigms for sustainability such as "water recycling and reusing".

Adra town in southern Spain was the case study chosen to show this problematic and introduce a feasible methodology to conceptualize the NSUS aquifer functioning and provide a tentative magnitude of the groundwater resource contributed to the SFUS. For this, findings from the geological, geophysical, hydrological, and hydrogeological surveys were combined to create a 1:5000 scale hydrogeological map and cross-sections, which are basic tools to design a proper urban water planning. The MASW and GPR geophysical techniques were especially useful for aquifer geometry definition. The NSUS average groundwater discharge was evaluated around 0.28 Mm^3 $year^{-1}$. Among other uses, this resource may guarantee a permanent water supply to watering public gardens and urban cleaning, which is currently around 0.11 Mm^3 $year^{-1}$ and will increase due to new urbanized areas, thus alleviating the pressure on the ARDGB.

This paper seeks to offer a feasible methodology for groundwater research in medium-size urban areas having steep topography, low-permeability bedrock underlying shallow urban aquifers, and peri-urban areas influencing the urban hydrology. The authors found that the introduced basic formulation for groundwater discharge enables for a tentative evaluation of this resource at most. For this reason, solving of the discussed hydrogeological gaps and designating of suitable sites for a sustainable groundwater exploitation regime will be subjects of future research.

Author Contributions: Conceptualization, F.J.A.; methodology, F.J.A., P.M.-P., M.C.P., and M.N.; formal analysis, P.M.-P., M.C.P., J.P.-C., and M.N.; data curation, M.C.P. and J.P.-C.; writing—original draft preparation, F.J.A.; writing—review and editing, F.J.A., P.M.-P., M.C.P., M.N., and F.D.; project administration, F.J.A.; funding acquisition, M.N. and F.D. All authors have read and agreed to the published version of the manuscript.

Funding: This research was partly supported by the Spanish Research Projects CGL2016-78075-P and CGL2007-66745-C02-02/BTE, and the Andalusian Research Projects IE17_5560_EEZA and P06-RNM-01732, all including European Regional Development Funds.

Institutional Review Board Statement: Not applicable.

Informed Consent Statement: Not applicable.

Data Availability Statement: The data presented in this study are available on request from the corresponding author.

Acknowledgments: The authors are grateful for the information on the urban and irrigation water supply systems provided by the Water Authority of Adra town and local Water Users Associations, respectively. Jorge M. Carvalho from University of Porto is also acknowledged for providing GPR equipment and software. The valuable comments and suggestions by two anonymous referees are greatly appreciated.

Conflicts of Interest: The authors declare no conflict of interest.

References

1. Sabater, S.; Barceló, D. *Water Scarcity in the Mediterranean. Perspectives Under Global Change*; Springer: Berlin/Heidelberg, Germany, 2010. [CrossRef]
2. Custodio, E. Coastal aquifers of Europe: An overview. *Hydrogeol. J.* **2010**, *18*, 269–280. [CrossRef]
3. Dalin, C.; Wada, Y.; Kastner, T.; Puma, M.J. Groundwater depletion embedded in international food trade. *Nature* **2017**, *543*, 700–704. [CrossRef] [PubMed]
4. Custodio, E.; Andreu-Rodes, J.M.; Aragón, R.; Estrela, T.; Ferrer, J.; García-Aróstegui, J.L.; Manzano, M.; Rodríguez-Hernández, L.; Sahuquillo, A.; Del Villar, A. Groundwater intensive use and mining in south-eastern peninsular Spain: Hydrogeological, economic and social aspects. *Sci. Total Environ.* **2016**, *559*, 302–316. [CrossRef] [PubMed]
5. Martínez-Valderrama, J.; Ibáñez, J.; Alcalá, F.J. AQUACOAST: A simulation tool to explore coastal groundwater and irrigation farming interactions. *Sci. Program.* **2020**, *2020*, 9092829. [CrossRef]
6. Alcalá, F.J.; Martínez-Valderrama, J.; Robles-Marín, P.; Guerrera, F.; Martín-Martín, M.; Raffaelli, G.; Tejera de León, J.; Asebriy, L. A hydrological-economic model for sustainable groundwater use in sparse-data drylands: Application to the Amtoudi Oasis in southern Morocco, northern Sahara. *Sci. Total Environ.* **2015**, *537*, 309–322. [CrossRef]
7. Alcalá, F.J.; Martín-Martín, M.; Guerrera, F.; Martínez-Valderrama, J.; Robles-Marín, P. A feasible methodology for groundwater resource modelling for sustainable use in sparse-data drylands: Application to the Amtoudi Oasis in the northern Sahara. *Sci. Total Environ.* **2018**, *630*, 1246–1257. [CrossRef]
8. Morote, Á.F.; Olcina, J.; Rico, A.M. Challenges and proposals for socio-ecological sustainability of the tagus–segura aqueduct (Spain) under climate change. *Sustainability* **2017**, *9*, 2058. [CrossRef]
9. Morote, Á.F.; Olcina, J.; Hernández, M. The use of non-conventional water resources as a means of adaptation to drought and climate change in Semi-Arid Regions: South-Eastern Spain. *Water* **2019**, *11*, 93. [CrossRef]
10. Licciardello, F.; Milani, M.; Consoli, S.; Pappalardo, N.; Barbagallo, S.; Cirelli, G. Wastewater tertiary treatment options to match reuse standards in agriculture. *Agric. Water Manag.* **2018**, *210*, 232–242. [CrossRef]
11. Allam, A.R.; Saaf, E.J.; Dawoud, M.A. Desalination of brackish groundwater in Egypt. *Desalination* **2002**, *152*, 19–26. [CrossRef]
12. Alcalá, F.J. Usefulness of the Cl/Br ratio to identify the effect of reverse osmosis treated waters on groundwater systems. *Desalination* **2019**, *470*, 114102. [CrossRef]
13. Downward, S.R.; Taylor, R. An assessment of Spain's Programa AGUA and its implications for sustainable water management in the province of Almería, southeast Spain. *J. Environ. Manag.* **2007**, *82*, 277–289. [CrossRef]
14. Rodríguez-Rodríguez, M.; Benavente, J.; Alcalá, F.J.; Paracuellos, M. Long-term water monitoring in two Mediterranean lagoons as an indicator of land-use changes and intense precipitation events (Adra, Southeastern Spain). *Estuar. Coast. Shelf Sci.* **2011**, *91*, 400–410. [CrossRef]
15. Alcalá, F.J.; Solé, A.; Creus, C.; Domingo, F. Aporte urbano de agua subterránea hacia masas hídricas regionales y ecosistemas dependientes. Caso de la localidad costera de Adra (SE de España). In *VII Simposio del Agua en Andalucía*, 1st ed.; López-Geta, J.A., Rubio, J.C., Martín-Machuca, M., Eds.; Geological Survey of Spain: Madrid, Spain, 2008; Volume 2, pp. 699–708.

16. Poeter, E.; Anderson, D. Multimodel ranking and inference in ground water modeling. *Groundwater* **2005**, *43*, 597–605. [CrossRef]
17. Hojberg, A.L.; Refsgaard, J.C. Model uncertainty-parameter uncertainty versus conceptual models. *Water Sci. Technol.* **2005**, *52*, 177–186. [CrossRef]
18. Beven, K. Towards integrated environmental models of everywhere: Uncertainty, data and modelling as a learning process. *Hydrol. Earth Syst. Sci.* **2007**, *11*, 460–467. [CrossRef]
19. Alcalá, F.J.; Espinosa, J.; Navarro, M.; Sánchez, F.J. Propuesta de división geológica de la localidad de Adra (provincia de Almería). Aplicación a la zonación sísmica. *Rev. Soc. Geológica España* **2002**, *15*, 55–66.
20. Monteiro Santos, F.A.; Sultan, S.A.; Represas, P.; El Sorady, A.L. Joint inversion of gravity and geoelectric data for groundwater and structural investigation: Application to the northwestern part of Sinai, Egypt. *Geophys. J. Int.* **2006**, *165*, 705–718. [CrossRef]
21. Khalil, M.A.; Hafez, M.A.; Santos, F.M.; Ramalho, E.C.; Mesbah, H.S.; El-Qady, G.M. An approach to estimate porosity and groundwater salinity by combined application of GPR and VES: A case study in the Nubian sandstone aquifer. *Near Surf. Geophys.* **2010**, *8*, 223–233. [CrossRef]
22. Alam, K.; Ahmad, N. Determination of aquifer geometry through geophysical methods: A case study from Quetta Valley, Pakistan. *Acta Geophys.* **2014**, *62*, 142–163. [CrossRef]
23. Farzamian, M.; Monteiro Santos, F.A.; Khalil, M.A. Estimation of unsaturated hydraulic parameters in sandstone using electrical resistivity tomography under a water injection test. *J. Appl. Geophys.* **2015**, *121*, 71–83. [CrossRef]
24. Binley, A.; Hubbard, S.S.; Huisman, J.A.; Revil, A.; Robinson, D.A.; Singha, K.; Slater, L.D. The emergence of hydrogeophysics for improved understanding of subsurface processes over multiple scales. *Water Resour. Res.* **2015**, *51*, 3837–3866. [CrossRef] [PubMed]
25. Gonçalves, R.; Farzamian, M.; Monteiro Santos, F.A.; Represas, P.; Mota Gomes, A.; Lobo de Pina, A.F.; Almeida, E.P. Application of time-domain electromagnetic method in investigating saltwater intrusion of Santiago Island (Cape Verde). *Pure Appl. Geophys.* **2017**, *174*, 4171–4182. [CrossRef]
26. Giustiniani, M.; Accaino, F.; Picotti, S.; Tinivella, U. Characterization of the shallow aquifers by high-resolution seismic data. *Geophys. Prospect.* **2008**, *56*, 655–666. [CrossRef]
27. Martorana, R.; Lombardo, L.; Messina, N.; Luzio, D. Integrated geophysical survey for 3D modelling of a coastal aquifer polluted by seawater. *Near Surf. Geophys.* **2014**, *12*, 45–59. [CrossRef]
28. Foti, S.; Hollender, F.; Garofalo, F.; Albarello, D.; Asten, M.; Bard, P.Y.; Comina, C.; Cornou, C.; Cox, B.; Giulio, G.D.; et al. Guidelines for the good practice of surface wave analysis: A product of the InterPACIFIC project. *Bull. Earthq. Eng.* **2018**, *16*, 2367–2420. [CrossRef]
29. Xia, J.; Miller, R.D.; Park, C.B. Estimation of near-surface shear-wave velocity by inversion of Rayleigh wave. *Geophysics* **1999**, *64*, 691–700. [CrossRef]
30. Xia, J.; Miller, R.D.; Park, C.B.; Hunter, J.A.; Harris, J.B.; Ivanov, J. Comparing shear-wave velocity profiles inverted from multichannel surface wave with borehole measurements. *Soil Dyn. Earthq. Eng.* **2002**, *22*, 181–190. [CrossRef]
31. Park, C.B.; Miller, R.D.; Xia, J. Multi-channel analysis of surface waves. *Geophysics* **1999**, *64*, 800–808. [CrossRef]
32. Park, C.B.; Miller, R.D.; Xia, J.; Ivanov, J. Multichannel analysis of surface waves (MASW)—Active and passive methods. *Lead Edge* **2007**, *26*, 60–64. [CrossRef]
33. Paz, C.; Alcalá, F.J.; Carvalho, J.M.; Ribeiro, L. Current uses of ground penetrating radar in groundwater-dependent ecosystems research. *Sci. Total Environ.* **2017**, *595*, 868–885. [CrossRef]
34. Bobeck, P. Henry Darcy in his own words. *Hydrogeol. J.* **2006**, *14*, 998–1004. [CrossRef]
35. Brown, G.O. Henry Darcy and the making of a law. *Water Resour. Res.* **2002**, *38*, 1106. [CrossRef]
36. Chen, D.; Chen, H.W. Using the Köppen classification to quantify climate variation and change: An example for 1901–2010. *Environ. Dev.* **2013**, *6*, 69–79. [CrossRef]
37. Herrera, S.; Cardoso, R.M.; Soares, P.M.M.; Espírito–Santo, F.; Viterbo, P.; Gutiérrez, J.M. Iberia01: A new gridded dataset of daily precipitation and temperatures over Iberia. *Earth Syst. Sci. Data* **2019**, *11*, 1947–1956. [CrossRef]
38. IGME. *Geological Map of Spain, Scale 1:50,000*; Sheet nº 1057 Adra. Geological Survey of Spain, Memory and Maps; IGME: Madrid, Spain, 1983; Available online: http://info.igme.es/cartografiadigital/geologica/Magna50.aspx (accessed on 11 January 2021).
39. Sanz de Galdeano, C. *La Zona Interna Bético-Rifeña. Antecedentes, Unidades Tectónicas, Correlaciones y Bosquejo de Reconstrucción Paleogeográfica*; University of Granada: Granada, Spain, 1997; pp. 1–316.
40. Alcalá, F.J.; Guerrera, F.; Martín-Martín, M.; Raffaelli, G.; Serrano, F. Geodynamic implications derived from Numidian-like distal turbidites deposited along the Internal-External Domain Boundary of the Betic Cordillera (S, Spain). *Terra Nova* **2013**, *25*, 119–129. [CrossRef]
41. Goy, J.L.; Zazo, C. Synthesis of the Quaternary in the Almería littoral neotectonic activity and its morphologic features, western Betics, Spain. *Tectonophysics* **1986**, *130*, 259–270. [CrossRef]
42. Pulido-Bosch, A.; Morales, G.; Benavente, J. Hidrogeología del delta del río Adra. *Estud. Geológicos* **1988**, *44*, 429–443. [CrossRef]
43. IGME. *Hydrogeological Map of Spain, Scale 1:200,000*; Sheet nº 84, Almería; Geological Survey of Spain, Memory and Maps; IGME: Madrid, Spain, 1988; Available online: http://info.igme.es/cartografiadigital/tematica/Hidrogeologico200.aspx (accessed on 11 January 2021).
44. IGME. *Hydrogeological Atlas of Andalusia. Geological Survey of Spain*; IGME: Madrid, Spain, 1998; Available online: https://aguas.igme.es/igme/publica/libros1_HR/libro110/lib110.htm (accessed on 11 January 2021).

45. INE. *Official Population Figures of Spanish Municipalities: Review of the Municipal Register—Population at 01 January 2019*; Spanish National Institute for Statistics: Madrid, Spain, 2019; Available online: https://www.ine.es/jaxiT3/Datos.htm?t=2857#!tabs-tabla (accessed on 29 December 2020).
46. JUNTA. *Hydrographic Demarcation of the Andalusian Mediterranean Basins—Appendix III: Uses and Demands*; Ministry of Agriculture, Fishery and Environment, Government of Andalusia: Sevilla, Spain, 2015; pp. 1–133.
47. CENTER. *Evaluation of the San Fernando Canal Irrigation Community. Municipal district of Adra (Almería)*; Ministry of Agriculture, Fishery and Feeding, Government of Spain: Madrid, Spain, 2007; pp. 1–161.
48. Alcalá, F.J.; Zapata, A.; Morillas-Arcos, M. Estructuras de detención/retención en la localidad de Adra (Almería). Posible alternativa al aprovechamiento de aguas pluviales. In *V Simposio Sobre el Agua en Andalucía*, 1st ed.; Pulido-Bosch, A., Pulido-Leboeuf, P.A., Vallejos-Izquierdo, A., Eds.; University of Almería: Almería, Spain, 2001; Volume 1, pp. 143–154.
49. ARWUA. *Groundwater Balance of the Adra River Alluvial Aquifer at Los Hurtados-Las Checas Place (Adra, Almería)*; Memory and Appendices; Adra River Water Users Association: Adra, Spain, 2007.
50. CSIC. *Scientific-Technical Advice to Study the Hydrogeological Ground Behavior on the Site and Neighbouring to the San Andrés III and San Andrés IV Buildings, Located at 12 and 14 Portón Street, in Adra Town (Almería)*; Memory and Appendices; Spanish National Research Council: Almería, Spain, 2010.
51. Benavente, J.; Castillo, A. Estudio Hidrogeoquímico de la cuenca del río Adra. *Estud. Geológicos* **1989**, *45*, 81–90. [CrossRef]
52. González-López, J. Water Resources Evaluation of the Adra River Valley Alluvial Aquifer, Almería Province. Bachelor's Thesis, Catholic University of Murcia, Murcia, Spain, 2015.
53. Martínez-Pagán, P.; Navarro, M.; Pérez-Cuevas, J.; Alcalá, F.J.; García-Jerez, A.; Vidal, F. Shear-wave velocity structure from MASW and SPAC methods. The case of Adra town, SE Spain. *Near Surf. Geophys.* **2018**, *16*, 356–371. [CrossRef]
54. Lunt, I.A.; Hubbard, S.S.; Rubin, Y. Soil moisture content estimation using ground penetrating radar reflection data. *J. Hydrol.* **2015**, *307*, 254–269. [CrossRef]
55. Annan, A.P. GPR—History, trends, and future developments. *Subsurf. Sens. Technol. Appl.* **2002**, *3*, 253–270. [CrossRef]
56. Beres, M.; Haeni, F.P. Application of ground-penetrating radar methods in hydrogeologic studies. *Ground Water* **1991**, *29*, 375–386. [CrossRef]
57. Cassidy, N.J. Electrical and magnetic properties of rocks, soils and fluids. In *Ground Penetrating Radar: Theory and Applications*, 1st ed.; Jol, H.M., Ed.; Elsevier: Amsterdam, The Netherlands, 2009; pp. 41–72.
58. Neal, A. Ground penetrating radar and its use in sedimentology: Principles, problems and progress. *Earth-Sci. Rev.* **2004**, *66*, 261–330. [CrossRef]
59. Bano, M.; Marquis, G.; Nivière, B.; Maurin, J.C.; Cushing, M. Investigating alluvial and tectonic features with ground-penetrating radar and analyzing diffractions patterns. *J. Appl. Geophys.* **2000**, *43*, 33–41. [CrossRef]
60. Paz, C.; Alcalá, F.J.; Ribeiro, L. Ground penetrating radar attenuation expressions in shallow groundwater research. *J. Environ. Eng. Geophys.* **2020**, *25*, 153–160. [CrossRef]
61. Van Dam, R.L.; Schlager, W. Identifying causes of ground-penetrating radar reflections using time-domain reflectometry and sedimentological analyses. *Sedimentology* **2000**, *47*, 435–449. [CrossRef]
62. Paz, M.C.; Alcalá, F.J.; Medeiros, A.; Martínez-Pagán, P.; Pérez-Cuevas, J.; Ribeiro, L. Integrated MASW and ERT imaging for geological definition of an unconfined alluvial aquifer sustaining a coastal groundwater-dependent ecosystem in southwest Portugal. *Appl. Sci.* **2020**, *10*, 5905. [CrossRef]
63. García-Jerez, A.; Navarro, M.; Alcalá, F.J.; Luzón, F.; Pérez-Ruiz, J.A.; Enomoto, T.; Vidal, F.; Ocaña, E. Shallow velocity structure using joint inversion of array and h/v spectral ratio of ambient noise: The case of Mula town (SE of Spain). *Soil Dyn. Earthq. Eng.* **2007**, *27*, 907–919. [CrossRef]
64. Mitchell, J.K.; Soga, K. *Fundamentals of Soil Behaviour*, 3rd ed.; Wiley: London, UK, 2005; pp. 1–592.
65. Zimmer, M.A.; Prasad, M.; Mavko, G.; Nur, A. Seismic velocities of unconsolidated sands: Part 1—Pressure trends from 0.1 to 20 MPa. *Geophysics* **2007**, *72*, E1–E13. [CrossRef]
66. McGann, C.R.; Bradley, B.A.; Cubrinovski, M. Investigation of shear wave velocity depth variability, site classification, and liquefaction vulnerability identification using a near-surface Vs model of Christchurch, New Zealand. *Soil Dyn. Earthq. Eng.* **2017**, *92*, 692–705. [CrossRef]
67. Navarro, M.; Vidal, F.; Enomoto, T.; Alcalá, F.; García-Jerez, A.; Sánchez, F.J.; Abeki, N. Analysis of the weightiness of site effects on reinforced concrete (RC) building seismic behavior. The Adra town example (SE Spain). *Earthq. Eng. Struct. Dyn.* **2007**, *36*, 1363–1383. [CrossRef]
68. Andreu, J.M.; Alcalá, F.J.; Vallejos, Á.; Pulido-Bosch, A. Recharge to aquifers in SE Spain: Different approaches and new challenges. *J. Arid Environ.* **2011**, *75*, 1262–1270. [CrossRef]
69. Martínez-Pagán, P.; Navarro, M.; Pérez-Cuevas, J.; Alcalá, F.J.; García-Jerez, A.; Sandoval-Castaño, S. Shear-wave velocity based seismic microzonation of Lorca city (SE Spain) from MASW analysis. *Near Surf. Geophys.* **2014**, *12*, 739–749. [CrossRef]
70. Lerner, D.N. Identifying and quantifying urban recharge: A review. *Hydrogeol. J.* **2002**, *10*, 143–152. [CrossRef]
71. Vázquez-Suñé, E.; Sánchez-Vila, X.; Carrera, J. Introductory review of specific factors influencing urban groundwater, an emerging branch of hydrogeology, with reference to Barcelona, Spain. *Hydrogeol. J.* **2004**, *13*, 522–533. [CrossRef]
72. Schirmer, M.; Leschik, S.; Musolff, A. Current research in urban hydrogeology—A review. *Adv. Water Resour.* **2013**, *51*, 280–291. [CrossRef]

73. Asebriy, L.; Cherkaoui, T.; El Amrani-El Hassani, I.; Franchi, R.; Guerrera, F.; Martín-Martín, M.; Guerrera-Patamìa, C.; Raffaelli, G.; Robles-Marín, P.; Tejera de León, J.; et al. Deterioration processes on archaeological sites of Chellah and Oudayas (world cultural heritage, Rabat, Morocco): Restoration test and recommendations. *Ital. J. Geosci.* **2009**, *128*, 157–171.
74. Luberti, G.M.; Vergar, F.; Marin, R.; Pica, A.; Del Monte, M. Anthropogenic modifications to the drainage network of Rome (Italy): The case study of the Aqua Mariana. *Alp. Mediterr. Quat.* **2018**, *31*, 119–132.
75. Forno, M.G.; De Luca, D.; Bonasera, M.; Bucci, A.; Gianotti, F.; Lasagna, M.; Lucchesi, S.; Pelizza, S.; Piana, F.; Taddia, G. Synthesis on the Turin subsoil stratigraphy and hydrogeology (NW Italy). *Alp. Mediterr. Quat.* **2018**, *31*, 147–170.
76. Alcalá, F.J.; Cantón, Y.; Contreras, S.; Were, A.; Serrano-Ortiz, P.; Puigdefábregas, J.; Solé-Benet, A.; Custodio, E.; Domingo, F. Diffuse and concentrated recharge evaluation using physical and tracer techniques: Results from a semiarid carbonate massif aquifer in southeast Spain. *Environ. Earth Sci.* **2011**, *62*, 541–557. [CrossRef]
77. Pulido-Velazquez, D.; Romero, J.; Collados-Lara, A.J.; Alcalá, F.J.; Fernández-Chacón, F.; Baena-Ruiz, L. Using the turnover time index to identify potential strategic groundwater resources to manage droughts within continental Spain. *Water* **2020**, *12*, 3281. [CrossRef]

applied sciences

MDPI

Article

Temporal and Spatial Groundwater Contamination Assessment Using Geophysical and Hydrochemical Methods: The Industrial Chemical Complex of Estarreja (Portugal) Case Study

Tiago Marques, Manuel Senos Matias, Eduardo Ferreira da Silva, Nuno Durães and Carla Patinha *

GEOBIOTEC Research Centre, Department of Geosciences, University of Aveiro, Campus de Santiago, 3810-193 Aveiro, Portugal; tiago.david@ua.pt (T.M.); mmatias@ua.pt (M.S.M.); eafsilva@ua.pt (E.F.d.S.); nunoduraes@ua.pt (N.D.)
* Correspondence: cpatinha@ua.pt

Abstract: With more than a half-century in operation, the industrial chemical complex of Estarreja (ICCE) in northern Portugal has left serious environmental liabilities in the region. Although protective measures were implemented, soils, surface, and groundwater contamination caused by persistent pollutants are still prevalent. This study presents data from several geophysical and hydrochemical campaigns carried out to monitor groundwater contamination in the Estarreja region over a period of 30 years. Both geophysical and hydrochemical data showed a good agreement and revealed an important anomaly caused by groundwater contamination (high levels of Na, Cl, SO_4, and Fe, among others) in 2006–2007, likely caused by the remobilization of waste pollutants (roasted pyrites, soils, and sludge) during their deposition in a sealed landfill (operating between 2003 and 2005). More recently, in 2016, this impact persists, but was more attenuated and showed a general migration pattern from E to SW according to one of the main groundwater flow paths. Groundwater flow in this region has a local radial behaviour. Drainage effluent systems, such as ditches and buried pipes formerly used by ICCE, are also likely to contribute to some contamination "hotspots". Finally, the results obtained by the combined use of these two approaches allowed for the delineation of the contamination plume for future monitoring.

Keywords: groundwater; contamination; time-space; geophysics; hydrochemistry

Citation: Marques, T.; Matias, M.S.; Silva, E.F.d.; Durães, N.; Patinha, C. Temporal and Spatial Groundwater Contamination Assessment Using Geophysical and Hydrochemical Methods: The Industrial Chemical Complex of Estarreja (Portugal) Case Study. *Appl. Sci.* **2021**, *11*, 6732. https://doi.org/10.3390/app11156732

Academic Editors: Francisco Javier Alcalá, Maria Catarina Paz, Pedro Martínez-Pagán and Fernando Monteiro Santos

Received: 14 June 2021
Accepted: 17 July 2021
Published: 22 July 2021

1. Introduction

Chemical industry can cause severe environmental damages, especially those that operated in the past when environmental protection measures were scarce or absent. This degradation is often not limited to the area next to the industries themselves, but instead is likely to affect a much larger area. This is particularly relevant when saturated media is affected, since its limited adsorption capability and continuous movement result in the subsurface migration and easy dispersion of contaminants.

In aquifers, particularly those formed by unconsolidated materials, which are highly porous and permeable, diffusion is the most important process in contaminant migration [1]. Once in the saturated media, pollutants move according to groundwater flow paths, unless physicochemical interactions with sediments affect their mobility.

Contamination effects can persist over long periods, even after their sources are isolated or removed. This can be instigated by back diffusion processes, i.e., the reverse diffusion of pollutants from low permeability layers (e.g., clayey formations, where contaminants were preferably adsorbed during high polluting water periods) to high permeability zones [2]. On the other hand, the persistence of a pollutant is also a factor of its intrinsic characteristics. Unlike several hazardous organic constituents, metal(loid)s cannot be degraded [3,4], while there are cases of non-biodegradable organic compounds [4] and others that exhibit half lives of hundreds to thousands of years, depending on the hydraulic behaviour [5]. The low microbial activity in groundwater is also a factor limiting biodegradation processes. Furthermore, the characteristics of many degradation products formed are still poorly understood and can present different reactivity under field conditions [5].

Although technical advances, environmental laws, and public awareness have imposed new procedures and legal restrictions to industry operations, the environmental legacy of the past is frequently difficult to overcome. That is why, in several cases, monitoring natural attenuation is the more reliable alternative to groundwater decontamination [5,6]. A complete study of groundwater contamination requires the investigation of larger areas (i.e., not only the immediate vicinity of known pollution sources), through the evaluation of relevant physical and chemical properties and parameters, as well as their variation in time throughout the region. In this sense, geophysical exploration methods have been applied with great success in the investigation of environmental problems [7,8]. These methods are particularly suited to produce fast and reliable spatial maps of physical properties related with groundwater contamination issues, but they are unable to identify the contaminant. Electrical resistivity or its inverse, electrical conductivity, is an important physical property and there are many examples and case studies in the literature regarding the use of resistivity-based methods for aquifer characterization. Typically, resistivity can vary from 1 to 10 for contaminated and pristine groundwater, respectively, [9] and, therefore, resistivity can be used to map contaminated areas.

Traditional resistivity methods using DC currents, or galvanic methods, have long been used for this purpose. At first, conventional resistivity soundings (VES) were used, but they provide 1D information [7]. Contamination plumes have a 3D character and the combination of traditional electrode arrays (Schlumberger, Wenner, dipole-dipole and others) along lines, electrical resistivity tomography (ERT), and grids provide 2D and 3D images of the areas under investigation [7,10]. However, resistivity methods imply physical contact with the ground, requiring the use of electrodes to pass current into the ground, and the setup of a complete field apparatus usually takes considerable time and effort. Furthermore, places with high contact or superficial resistivity, such as dry conditions, sand, rock outcrops, and paved areas, will cause additional field problems.

Alternatively, electromagnetic methods using AC currents, or induction methods, do not require physical contact with the ground and measurements can be done as the operator walks. These methods can also be adapted for airborne use [10]. They are fast and easy to operate and can provide 2D and 3D images of the area under investigation with less field effort. However, if there are sources of electrical noise in the area, such as power lines, they can provide unstable, if not completely improper, data.

Groundwater contamination is a dynamic phenomenon. Thus, if reliable information is to be obtained, a 4D approach, namely, time-lapse studies, are required.

The use of time-lapse resistivity in groundwater contamination, as well as algorithms for inversion and model resistivity data, have long been discussed [11,12]. These problems have been recently addressed by a number of authors [13–17].

However, if data from different epochs are to be used, they must be compared and normalized to a known background, so that resistivity variations can be attributed solely to the variation of groundwater contamination.

The apparent electrical conductivity of sediments and soil layers is complex due to the interaction of several factors, such as pore-water salinity, water content, and particle size distribution [18]. Hydrochemical parameters are a useful and complementary tool for assessing environmental aquifer conditions by providing information about chemical signatures of groundwater, distinguishing geological formations with similar hydrogeochemical characteristics, and understanding element mobility and dispersion patterns [19].

Therefore, the use of both electrical exploration methods, to delimit contaminated areas as well as to estimate their temporal and spatial evolution, and hydrochemical parameters, to access the level and type of groundwater contamination, have been applied with success in several studies [20–23].

The industrial chemical complex of Estarreja (ICCE) in northern Portugal has been operating for almost 70 years. During this period, environmental laws have evolved from non-existent to strict regulations in accordance with the EU regulations. Thus, in the past, effluents were discharged in local streams, man-made ditches, and buried metal pipes, which were easily corroded, playing a major role in the contamination of local aquifers and soils. Approximately 70% of the population living near the ICCE has water wells that are widely used for agriculture (~81%) and domestic (about 24% for drinking) supplies [24]. For this reason, the understanding of the spatiotemporal groundwater contamination is of paramount interest for the population, as well as for authorities and technicians in charge of remediation operations.

Since there is a large volume of geophysical and hydrochemical data acquired over several decades in the region, this study integrated such information to investigate temporal (a time span of about 30 years) groundwater contamination in the ICCE region in order to:

(a) Map contamination plume(s) and evaluate its/their progression;
(b) Estimate the propagation velocity of plume(s) and its/their variation during the study period;
(c) Investigate the sources of contamination and their evolution (type and intensity);
(d) Carry out an assessment of the remediation works implemented in the early 2000s;
(e) Evaluate the benefits of this joint approach in future monitoring and remediation operations, so that the findings of this work can have a wider use and applicability to other case studies.

2. Site Characterisation

2.1. The Industrial Study Site

The industrial chemical complex of Estarreja (ICCE) occupies an area of approximately 2 km^2 (Figure 1), and is located 1 km east of Estarreja and 20 km north of Aveiro city in northern Portugal.

This industrial complex started operations in the 1950s and, since then, has gone through different production cycles. The more relevant products have been [25]: ammonium sulphate (1952–1990s) manufactured from sulphuric acid (obtained by pyrite roasting) and ammonia; nitric acid and ammonium nitrate (1974–1990s); nitric acid, aniline, and nitrobenzene (1978–present); sodium and chlorate compounds using mercury cathodes (1956–2002); polyvinyl chloride—PVC (1963–1983); and isocyanide polymers of aromatic base (1978–present). Consequently, large amounts of solid wastes, some enriched in metal(loid)s (e.g., As, Cu, Pb, Hg, and Zn) were produced, which were latter deposited in a sealed landfill, while others were deposited in a calcium hydroxide mud dam (Figure 1). In addition, effluents are also responsible for a high pollution load containing organic compounds (namely, aniline, benzene, monochlorobenzene, and mononitrobenzene) as well as metal(loid)s (e.g., As, Hg, Pb and Zn). Until 1975, these effluents were transported by artificial (open and permeable) ditches (Vala de S. Filipe, Vala da Breja, and Vala do Canedo), which discharged into the natural surface water streams [25]. In the present, they are transported by an emissary pipe to a water treatment plant and only then discharged into natural water systems.

The main environmental remediation interventions were undertaken from 1994 onwards by a joint effort of the local council, industries, and national environmental authorities. Since that time, chemical plants have also adopted procedures to control the emission of effluents and solid wastes to reduce their environmental impacts. Examples of the interventions carried out at the ICCE include the replacement of Hg cells with membrane cells in 2002 and the containment of about 300,000 tons of waste and contaminated soils in a confined landfill (2003–2005), supported by the ERASE project, an association between industry and local authorities [26].

Groundwater contamination studies in the ICCE started in the last quarter of the 20th century [27–29], and since then many other works have been carried out, devoted either to the environmental impacts [25,30–34] or to the potential associated health risks [35–37].

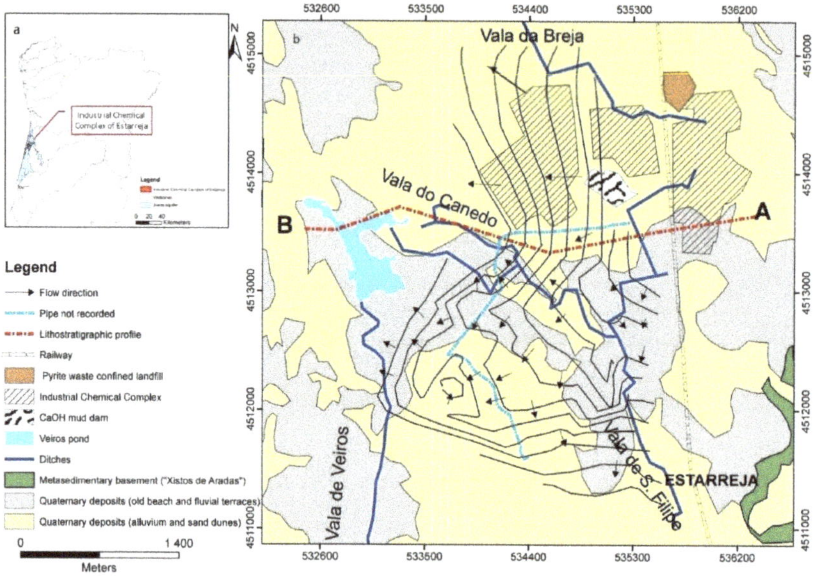

Figure 1. (**a**) Location of the Aveiro aquifer boundary and the industrial chemical complex of Estarreja, northern Portugal; (**b**) Simplified geological map of the study area (adapted from Teixeira and Zbyszewski [38]) and main groundwater flow paths (compiled data from measurements made in 2015 and from Ordens [33]); map coordinates: UTM; WGS84—29N zone.

However, in parallel to the intense industrial activity, agriculture and farming have also been intensive and must be accounted for in terms of the contamination process in the area. In fact, these activities exert further pressure on local ecosystems because of the widespread use of agrochemicals and the effluents from cattle breeding.

2.2. Geology and Hydrogeology

The study area is part of the so-called Aveiro sedimentary basin, located in the northern sector of the western Meso-Cenozoic Margin. The stratigraphic sequence of this basin is discontinuous, consisting mainly of Holocene-Pleistocene units and, with less representation, Cretaceous and Triassic formations. These sedimentary units are deposited on a substrate of Neoproterozoic age, the Aradas shales formation, mainly composed of fine-grained metapelites of low metamorphic grade, which constitutes the Hercynian basement (Figure 1) [38].

In detail, the sedimentary sequences (from bottom to top) begin with a red-coloured polygenic conglomerate superimposed by Triassic fine and red sandstones. The Cretaceous units show a greater development towards the Atlantic coast and are deposited over Triassic materials or directly on the Hercynian bedrock. The predominant lithology of this sequence are sandstones with depositional characteristics that vary in depth, allowing for the individualization of different formations. Thus, the base of the sequence is composed of sandstones of variable grain size interbedded with clays and superimposed by a thick clayey layer. Covering this there is a white sandstones sequence, interspersed with clays and conglomerates, and an uppermost layer containing sandstones and clay-silt lagoon facies. Finally, the Quaternary deposits include Pleistocene beach and fluvial deposits related to two sedimentary cycles: (a) the lower unit shows a coarsening downward sequence, with coarse sands and pebbles at the base; and (b) the upper sequence is composed mainly of organic muds, clays, and clayey sands. The most recent Holocene sedimentary units include alluvium and sand-dunes.

Based on the texture and structure of different sedimentary sequences, two main aquifer systems are defined in this region: the Aveiro Cretaceous confined multilayer aquifer system, and the Aveiro Quaternary unconfined (locally semi-confined) multilayer aquifer system [39]. In the studied area, only the Aveiro Quaternary aquifer system is of interest, since the Cretaceous sequences are only expressed to the west of this location (Figure 2). The Aveiro Quaternary aquifer system comprises of three hydrogeological units [39]. The uppermost unit consists of a Holocene shallow unconfined aquifer with thickness varying between 8 and 10 m. This aquifer is highly permeable and shows a main flow path from east to west. Below this exists a semi-confined aquifer, composed of coarse and permeable sediments (sand and gravel), with groundwater flowing from east to west. The top of this aquifer consists of a low permeability mud and fine silt layer that constitutes the top of the Plio-Pleistocene sequence. Lastly is a semi-confined aquifer, formed by old beach and alluvial coarse, permeable sediments (sands, gravel, and pebbles), with groundwater flowing from west to east in opposition to the previous aquifers.

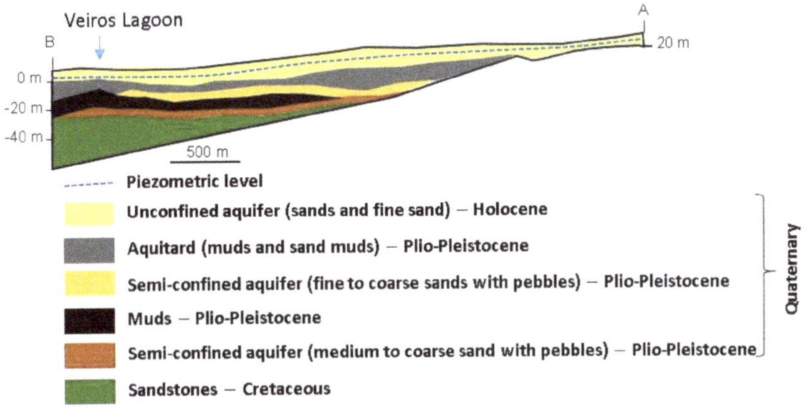

Figure 2. Simplified vertical profile (section AB marked in Figure 1) of the main units of the Quaternary aquifer system in the Estarreja region (adapted [33]). The vertical-to-horizontal scale ratio is 10:1.

It should be noted that recharge of this aquifer system is done mainly by rainfall infiltration and irrigation return. In Estarreja, the annual average precipitation is about 1050 mm, the annual average evapotranspiration is 650 mm, and the annual average bulk runoff (including surface runoff and aquifer recharge) is 400 mm [40]. The high permeability and porosity of the aquifer layers favours recharge and the groundwater flow gradient, but this also represents a potential contamination risk since the ICCE is in direct contact with the unconfined aquifer, which facilitates contaminants migration.

3. Materials and Methods

3.1. Antecedents

Over a period of almost 30 years, several geophysical and hydrochemical surveys, aimed at studying contamination and pollutants dispersion in the ICCE, were conducted in the Estarreja region.

Geophysical mapping of the area to investigate groundwater contamination started in 1975 [27] and continued later [31], but the density of observations was low and is not suitable to compare with later, denser observations or with hydrochemical data. Therefore, in order to obtain the best temporal and spatial representation of groundwater contamination in the ICCE area, several geophysical and hydrochemical campaigns were selected. The campaigns were those that best fit in terms of the area covered and those representing the pre- and post-environmental remediation operation periods, that is, before and after 2003–2005.

3.2. Geophysical Survey

The use of electromagnetic methods (EM) in the study of environmental problems have proved very suitable at identifying the boundaries of contamination plumes and their transport pathways in the saturated zone. There are numerous equipment and field strategies that can be used to carry out such an investigation [41]. Since the studied area has about 25 km^2, EM methods, being fast and reliable, were selected to map the electrical conductivity. The extensively tested EM34 was used by repeating the measurements in the same position and measuring both vertical fields (VF) with horizontal coils, and horizontal fields (HF) with vertical coils. Measurements were taken along local pathways with a 20 m interval, or inter coil spacing.

Over the period of ICCE area monitoring, several geophysical field surveys were conducted via resistivity, frequency domain electromagnetics, seismic refraction, and ground probing radar methods. In this work, frequency domain measurements from three surveys, Taunt [42], Ordens [33], and Marques [34], were chosen to evaluate the electrical conductivity evolution in the area surrounding the ICCE, and to compare with the hydrochemical findings gained in field campaigns. The choice of these surveys is justified as they covered the whole area in a very similar way. Furthermore, in 2016, two EM campaigns were carried out, one at the end of the wet season (February) and another at the end of the dry season (September) to investigate seasonal changes caused by the precipitation regime and subsequent groundwater level oscillations.

3.3. Hydrochemical Survey

Three groundwater sampling campaigns (1989, 2007, and 2016), carried out in the same area as that of the geophysical surveys, were considered in this study. Groundwater hydrochemical data of the selected campaigns were obtained from Barradas [43], Ordens [33], and Marques [34] in 1989, 2007, and 2016, respectively. Groundwater samples were collected in both holes and pumping wells. It should be noted that some of these wells are now abandoned due to unsafe levels of contamination.

The physicochemical parameters (temperature, electrical conductivity, and pH) were measured in situ using specific electrodes for each parameter. For chemical analysis, the water samples were filtered through a 0.45 μm pore size filter and stored in high-density polyethylene bottles, which were kept refrigerated in coolers during transport to the laboratory.

The analyses of chloride (Cl^-) and sulphate (SO_4^{2-}) were done by ion chromatography, while major cations and trace elements were determined by atomic absorption spectrometry (after a pre-concentration of waters by lyophilisation) for the 1989 sampling campaign, and by inductively coupled plasma mass spectrometry (ICP-MS) for the samples collected in 2007 and 2016.

4. Results and Discussion

4.1. Geophysical Survey

4.1.1. Influence of Seasonal Variations on Geophysical Data

The rainfall distribution in the region includes a rainy season from November to March (values over 150 mm/month) and a dry season from July to September (values less than 20 mm/month) [40]. These data, in conjunction with the high porosity and permeability of shallow geological formations, induce changes in the unconfined aquifer storage and groundwater quality. Therefore, it was necessary to investigate seasonal effects in shallow geophysical measurements if readings taken at different times of the year are to be compared.

The electrical conductivity measurements for two profiles (1 and 2; Figure 3A) obtained in the dry (September 2016) and in the wet (February 2016) seasons at sites with distinct features (south of the mud dam and close to the ditch Vala da Breja, respectively) and with different directions (W-E and NW-SE, respectively) are shown in Figure 3B.

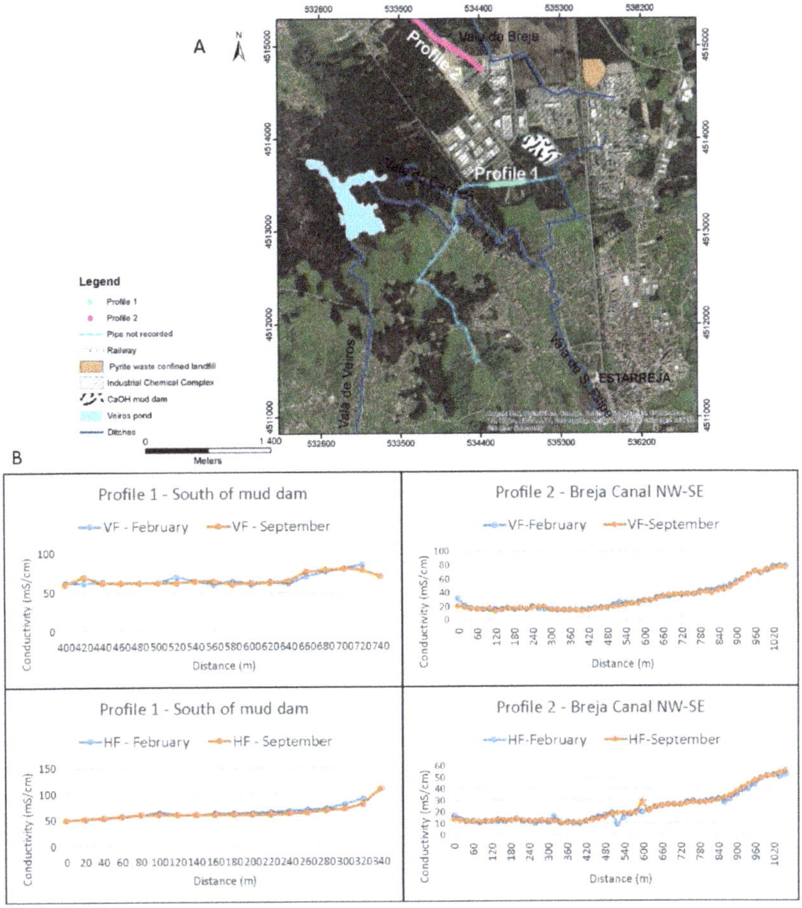

Figure 3. (**A**) Location of profiles 1 and 2 (map coordinates: UTM; WGS84—29N zone); (**B**) Seasonal variation of electrical conductivity in profiles 1 and 2 in the dry (September) and wet (February) season of 2016 (HF—horizontal fields; VF—vertical fields).

Since the reported estimated depths of exploration for the vertical and horizontal coils surveys were 15 m and 30 m, respectively [44], data from the vertical coils survey was expected to be influenced by the unconfined aquifer, while data obtained from the horizontal coils survey should be related to a deeper area, including the semi-confined aquifer (Figure 2).

According to the data, no striking differences between the surveys carried out in February and September of 2016 were obtained, proving that geophysical data are not markedly influenced by seasonal variations in this area.

4.1.2. Temporal Variation of Conductivity in the Area

Considering that seasonal changes in the geophysical data are not significant for the EM34 geometry used, the spatial variations of electrical conductivity for the area in the vicinity of ICCE were compared for 2001, 2006, and 2016 (Figure 4) regardless of the period of the year the measurements were taken. As observed, for all the periods considered, the highest conductivity values were found near the ICCE with an overall gentle decrease as the distance from this site increases. In general, these values distribution over the area replicate the main groundwater flow paths (i.e., E-W and E-SW).

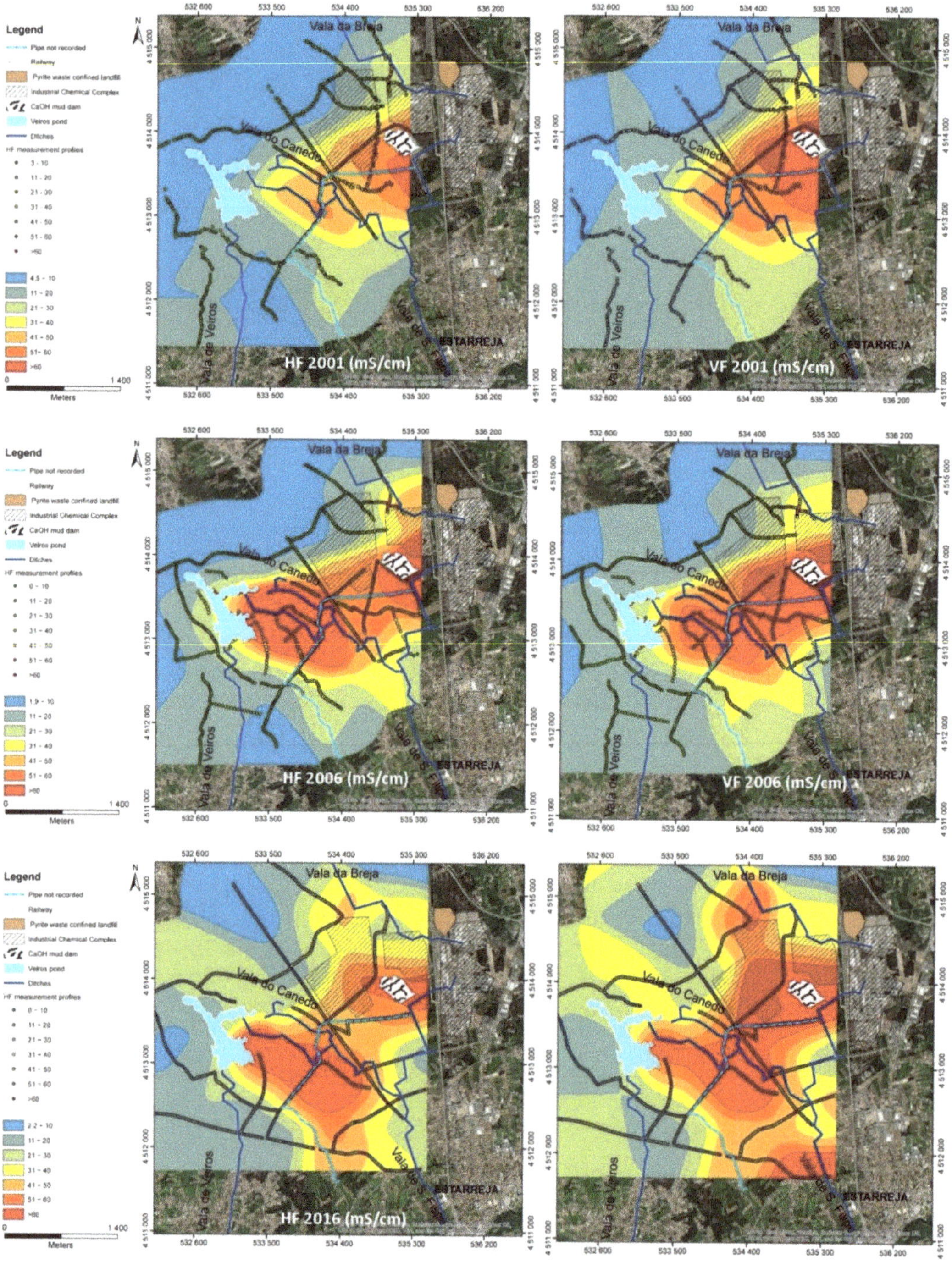

Figure 4. Electrical conductivity maps from 2001, 2006, and 2016 obtained through electromagnetic (EM34) methods. (HF—horizontal fields on the left; VF—vertical fields on the right). Map coordinates: UTM; WGS84—29N zone.

In addition, data measured with vertical coils (referring to shallower investigation depths, corresponding to the unsaturated and saturated zones of the unconfined aquifer) and data obtained with horizontal coils (reaching deeper depths of the unconfined and semi-confined aquifers) showed similar locations for the main electrical conductivity hotspots (Figure 4). Thus, the most important hotspots were located: (a) in the vicinity of the mud dam (Figure 4, area A), at first to the southwest (maps representing 2001 data), and then to the west and south; (b) the region to the east of the Veiros pond (Figure 4, area B on the maps from 2006 and 2016), whose anomaly later extends to the east, towards the anomaly of the mud dam generating a large contaminated area; (c) anomalies located near the ditches that carried effluents away from the ICCE, (Figure 4, area C on the bottom maps); and (d) a linear anomaly to the south, located over the expected position of an old buried pipe also used to carry effluents away from the ICCE (Figure 4, area D on the maps representing data from 2016). Another important feature on the maps of Figure 4 is the low intensity anomaly just to the west of the sealed landfill. This anomaly is very weak in the earlier maps, but increases in intensity in the central maps (Figure 4, area E on the 2006 maps). Afterwards, this anomaly showed an evident decrease, both in intensity and spatially, proving the efficiency of the remediation works.

The differences in the hotspots for vertical coils (Figure 4; maps on the left) and horizontal coils (Figure 4; maps on the right), mostly refer to the intensity values. The horizontal fields showed lower conductivity values due to the greater relative representation that the non-saturated zone has at the depths reached by the vertical coils. Notwithstanding, both HF and VF maps showed that the unconfined and the semi-confined aquifers were impacted, which was particularly evident in the measurements for 2006 and 2016, and clearly revealed the extension and evolution of the contamination plume, providing a general qualitative image of the impacts near the ICCE. From 2006 onwards, data revealed remarkably high conductivity values just to the east of the Veiros pond, which were not observed in 2001, and thus must correspond to some event that subsequently occurred.

As several remediation actions took place and industrial processes altered over time, it is necessary to investigate the effects, if any, of these measures in the contamination process of the area.

The information provided by the maps in Figure 4 is qualitative and reveals contaminated areas, hotspots, and their general geographic extension. However, it is difficult to extract quantitative information concerning contamination spread. Therefore, further data analysis is required if detailed information on preferential paths, velocities, and contamination spread are to be obtained.

Locally, groundwater roughly follows a divergent radial flow from the ICCE to the west (Figure 1; [33]). In addition, to the east of the ICCE, the thickness of the Quaternary formations decreases (Figure 2), which means that the importance of the unconfined aquifer is limited and there are no pumping wells. Therefore, a good approach to investigate temporal and spatial variations of geophysical data can be accomplished by gathering information along the lines forming a radial network from the ICCE, as shown in Figure 5. In this figure, the yellow lines depict the directions from N30° W to N120° W (with a 15° angular spacing) used to evaluate the variation of the measured conductivity values. The E-W line was not included because it runs into the Veiros pond.

For two orientations, N75° W and N105° W, the spatiotemporal variations correspond to the areas depicting the highest conductivity values, as shown in Figure 6. This figure was constructed by plotting field conductivity values in accordance with the time of measurement and the distance from the beginning of the profiles near the ICCE. These two orientations were chosen to illustrate the procedure, but they also correspond to important directions of contamination propagation, which was obtained by crossing information from Figure 4 with the main local radial flow directions of groundwater presented in Figure 1.

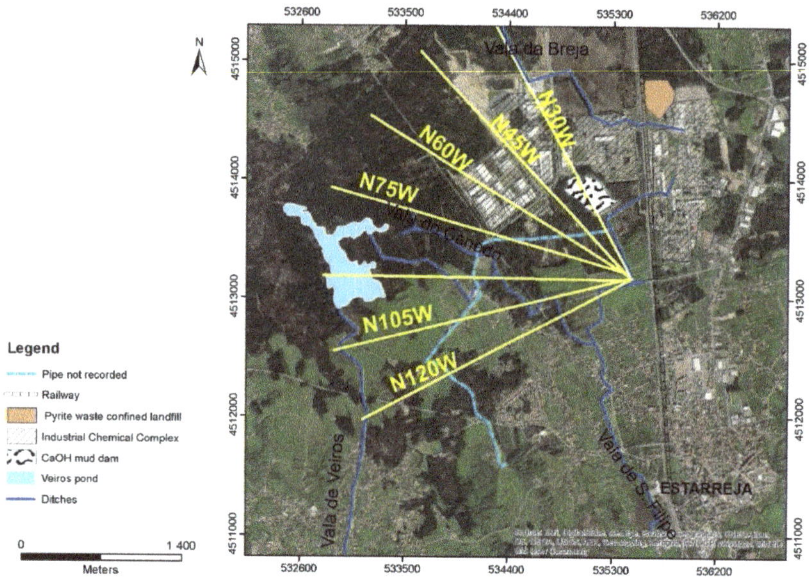

Figure 5. Transects used to sample the shift of the conductivity contouring lines (map coordinates: UTM; WGS84—29N zone).

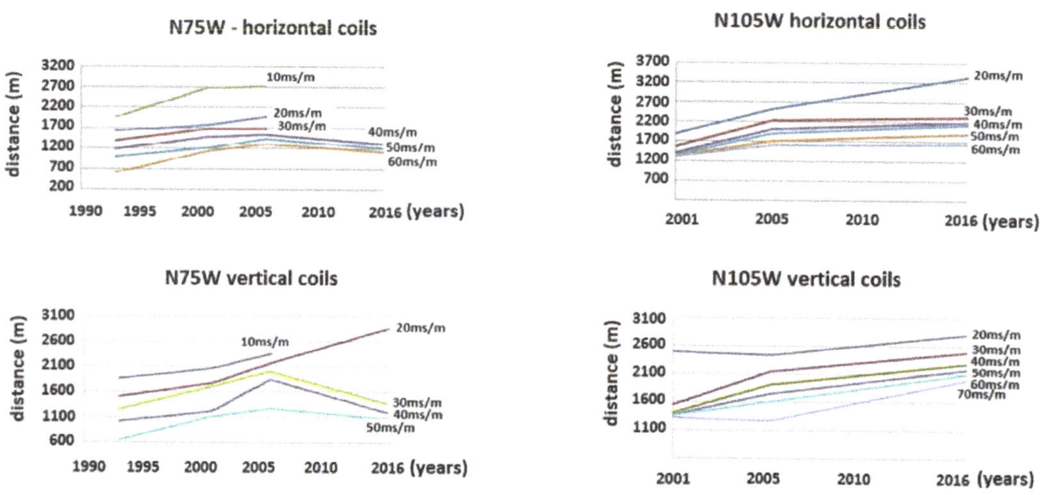

Figure 6. Electrical conductivity variation along profiles N75° W and N105° W over time.

Data from orientation N75° W, on the left of Figure 6, showed an increase of conductivity with distance until 2006. Afterwards, VF data (Figure 6, top left) depicted a gradual decrease in conductivity values and, hence, an expected decrease in contamination as a result of the implemented remediation works. For HF (Figure 6, bottom left), or shallower information, a similar behaviour was observed, in particular for the higher conductive curves. Hence, it seemed that the contamination also decreased at shallower depths, but there were still some signs of progression at lower levels of conductivity. This could be explained by surface runoff or very shallow contamination from an origin other than industrial sources. Therefore, for this orientation, the 2003–2005 remediation works appear to have been effective.

On the right of Figure 6, data from orientation N105° W are depicted. In this case, for both horizontal and vertical fields, contamination is shown to progress as distance increases in both graphs. However, there is a distinct change in the dip of the lines. This dip is gentler after 2006, revealing a lower propagation velocity of contamination, both for shallower and deeper areas. Therefore, remediation works also showed some degree of efficiency, although lower than that observed for N75° W.

Data from all the orientations in Figure 5 are summarized in Table 1. Data analysis showed similar behaviour for both shallower and deeper investigations, suggesting there is a hydraulic connection between the upper unconfined aquifer and the deeper semi-confined aquifer. This can be explained by discontinuous mud layers or their disturbing by human intervention, such as through the opening of wells or boreholes.

Table 1. Calculated dislocation of the conductivity (σ) contouring lines for vertical fields (VF) and horizontal (HF).

VF—Vertical Fields															
	1994–2001			2001–2006						2006–2016					
σ S/m	N60° W	N75° W	N120° W	N30° W	N45° W	N60° W	N75° W	N105° W	N120° W	N30° W	N45° W	N60° W	N75° W	N105° W	N120° W
10	−11	86		10	5	4	12					28			
20	−10	18	1	24	1	1	38	127	33		91	31		82	19
30	−10	35	14	14	2	1	5	136	25		77	35		8	31
40	−12	36		5	1	1	12	121	70		68	28	−22	19	22
50	−15	27		−12	−3	−1	42	110	129	21	63	30	−22	22	10
60	−20	66		−29	−12	−2	35	75		29	62	31	−18	20	
70				−44	1			59		35	51			4	
80				−55						39					
90				−62											
HF—Horizontal Fields															
	1994–2001			2001–2006						2006–2016					
σ S/m	N60° W	N75° W	N120° W	N30° W	N45° W	N60° W	N75° W	N105° W	N120° W	N30° W	N45° W	N60° W	N75° W	N105° W	N120° W
10	4	25		29	9	−1	59					36			
20	3	33	8	10	−1	−3	80	−12	35		77	40	69	38	23
30	−3	57		39	−1	−5	59	122	44	6	64	19	−63	34	30
40	−11	25		−12	3	−12	126	100	130	14	22	12	−64	38	38
50	−21	58		−9	17	−37	32	77		19	13	18	−20	43	
60				−3	9			52		12	11			48	
70								48		−2				45	
80								−10						72	

Notes: (1) propagation velocities are presented in metres/year; (2) negative values corresponding to a decrease in local contamination.

For particular directions, there was a difference before and after 2006, although the behaviour varies with orientation.

Thus, until 2001, there was a general decrease in curve displacements velocity in the direction N60° W, whereas for direction N75° W there was a general increase in contamination spread for both horizontal and vertical coils. Profile N120° W depicts residual displacements. Therefore, for this period, it seemed contamination receded to the northwest, but increased to the west of the mud dam. Geophysical data available for this period does not allow for any further discussion.

For 2001–2006, geophysical data provided a better scenario as measurements covered the area comprehensively. For this period, contamination spread was shown to recede for directions N30° W, N45° W, and N60° W, that is, the northern part of the area. This is particularly evident for the higher conductivity curves. Thus, if high conductivity values are attributed to contamination, it is likely that contamination receded during this period in this region. On the other hand, directions N75° W, N105° W, and N120° W showed a general increase in contamination spreading, thus the highest velocities were estimated and, furthermore, spreading affected all the conductivity curves in this region. In particular, values for horizontal coils, or data containing information from the deeper aquifer, depicted the highest calculated dislocations of conductivity curves.

From 2006 onwards, there was an increase in directions N30° W, N45° W, and N60° W, possibly in association with the Vala da Breja ditch, which is in the northern part of the region. For all the other directions, for example to the west and southwest, spreading either receded or the calculated velocity decreased significantly, possibly because of the remediation works carried out in the 2003–2005 period.

4.2. Hydrochemical Surveys and Water Quality

The most relevant physicochemical parameter indicators of the main inorganic contamination processes, caused by ICCE in the groundwater from the topmost layers of the Aveiro Quaternary aquifer system (i.e., the unconfined aquifer), are shown in Table 2. As observed, mean and median values for the analysed parameters were considerable higher in the 2007 sampling campaign than in the others. The exception was pH, which exhibited a slight decrease in the 2007 campaign. It should be noted that the 2007 campaign also showed the highest data variation for several parameters such as pH (4.4–10.1), electrical conductivity (EC; 50–26,800 μS cm^{-1}), Cl (8–14,840 mg L^{-1}), Na (6.1–10,600 mg L^{-1}), and Fe (10–167,000 μg L^{-1}). Sulphate was the exception, showing an increase over the study period, with a maximum value (3037 mg L^{-1}) in the 2016 campaign (Table 2).

Table 2. For the selected periods (1989, 2007, and 2016), summary of statistics for analytical data and groundwater samples in the study area.

Parameter	Campaign Year	Unit	Minimum	Mean	Median	Maximum	Standard Deviation	Samples
EC	1989	μS cm^{-1}	143	574	470	1785	350	60
	2007		50	3849	1102	26,800	5812	34
	2016		102	1687	613	13230	3087	32
pH	1989		3.8	6.0	6.2	7.3	0.69	60
	2007		4.4	5.5	5.3	10.1	1.04	34
	2016		3.5	6.0	6.3	7.8	0.97	32
Cl	1989	mg L^{-1}	17	89.65	67.5	237	59.88	60
	2007		8	1268.28	183.0	14,840	2735.06	34
	2016		17	387.44	72.6	2903	851.20	32
SO_4	1989	mg L^{-1}	10	82.53	69.0	297	56.20	60
	2007		15	364.07	147.0	2130	507.06	34
	2016		13	232.88	60.5	3037	567.26	32
Na	1989	mg L^{-1}	8.0	56.0	46.0	145	38.4	60
	2007		6.1	920.5	146.5	10,600.0	1984.7	34
	2016		13.9	288.5	72.0	2366.8	615.3	32
Fe	1989	μg L^{-1}	65.0	592.6	162.5	11,000.0	1609.9	60
	2007		10.0	9922.6	960.0	167,000.0	29,047.8	34
	2016		3.6	144.3	21.0	1458.9	343.3	32

In general, the first campaign (1989) showed the lowest mean electrical conductivity value (574 μS cm^{-1}), as well as the lowest concentration of the analysed ions (Cl^- 89.65 mg L^{-1}; SO_4^{2-} 82.53 mg L^{-1}; and Na^+ 56.0 mg L^{-1}). However, for Fe the highest mean value (9922.6 μg L^{-1}) was recorded in the 2007 campaign. For the 1989 and 2016 campaigns, the mean and median pH values were almost the same (Table 2).

As expected, the EC and concentration variations of the analysed parameters were positively correlated, while pH values were inversely correlated. In addition, the temporal evolution of these parameters was in accordance with that obtained in the geophysical surveys, which showed a substantial increase in the spread of contamination in the period of 2001–2006, and a general attenuation of electrical conductivity from then onwards. However, there were new hotspots that emerged at different sites (mainly associated with deeper aquifer measurement or vertical fields), as depicted in Figure 6 and Table 1.

Considering that groundwater EC is the parameter that best reflects contamination, a spatial distribution map of this parameter, according to the range of values and including the data from the three campaigns, is presented in Figure 7. As can be observed, the areas showing higher EC values were located SW of ICCE, which is particularly evident for 2007 and 2016, and in agreement with the geophysical data (Figure 4). In general, the contamination plume was likely to have migrated from E to the SW, away from the ICCE, along one of the main groundwater flow paths, but it also showed some attenuation effects.

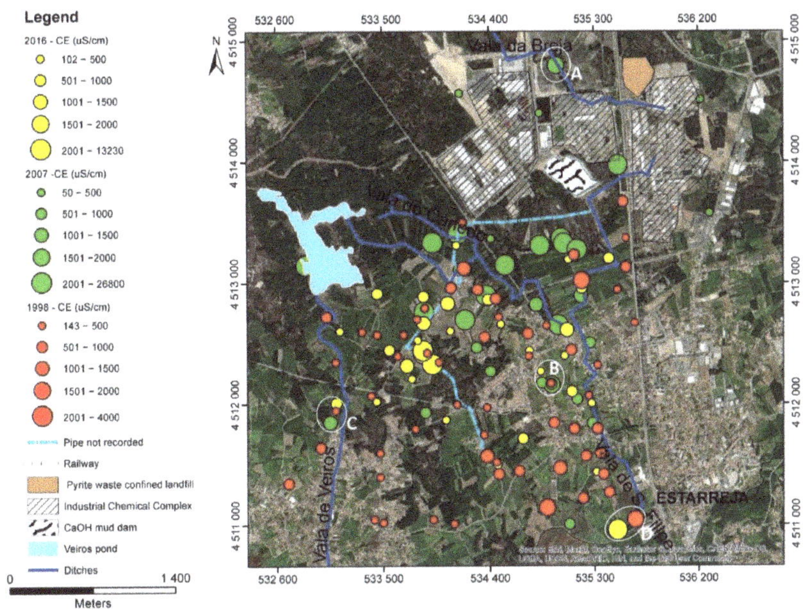

Figure 7. Spatial distribution of electrical conductivity values in the studied area (map coordinates: UTM; WGS84—29N zone).

This suggests that contamination spreads by diffusion because of the main geological features hosting the aquifer system, namely, highly porous and permeable media. Nevertheless, significant changes in the EC values between nearby locations also occurred for measurements carried out in the 2007 and 2016 campaigns, which reinforces the combination of the local divergent radial flow (Figure 1) with local superficial contamination hotspots (Figure 7; A, B, and C sites). In addition, high EC values were also found to the south of the sampling area, particularly in 1989 and 2016, closer to the end portion of the industrial draining ditches (Figure 7, area D).

Considering the most important plume propagation path (E to SW), obtained by both geophysical (Figure 4) and hydrochemical approaches (Figure 7), a group of samples from each campaign along that path (Figure 8) were selected to illustrate the variation of hydrochemical parameters, either in terms of time or distance from the ICCE (Figure 9).

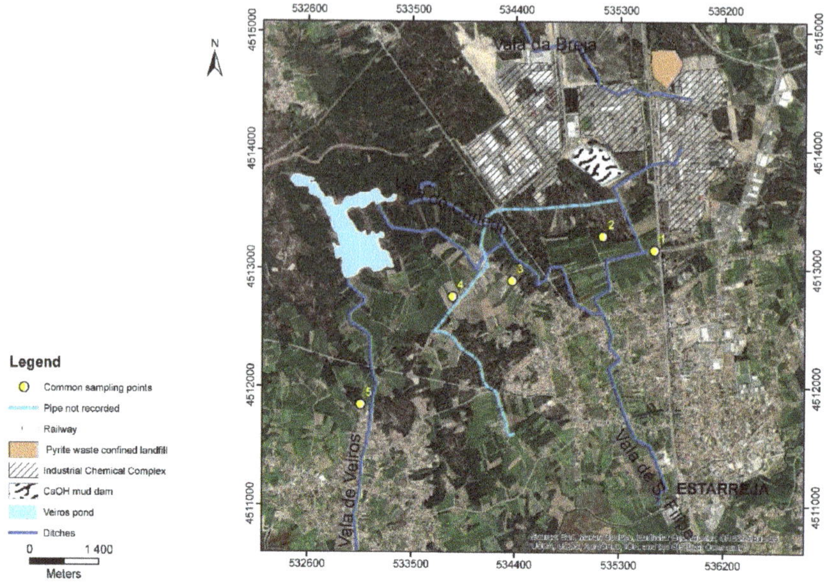

Figure 8. Points defined by groups of samples in common for the three sampling campaigns, illustrating the transect of the main plume (E-SW) alignment (map coordinates: UTM; WGS84—29N zone).

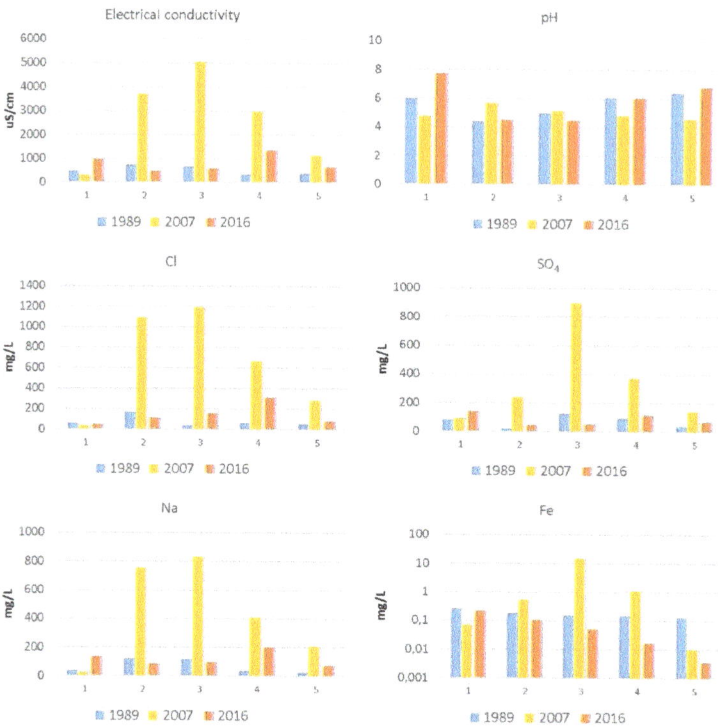

Figure 9. For the 1989, 2007, and 2016 campaigns, variation of electric conductivity (EC), pH, Cl, Na, SO_4, and Fe along the E-SW profile (displayed in Figure 8).

In Figure 9, each bar represents the element concentrations in the three different campaigns. Electrical conductivity, Cl, Na, and SO_4 values showed a similar spatial trend for each sampling campaign, revealing that Cl, Na (likely related to brines), and SO_4 (related to acid sulphuric production) were responsible for the EC changes. This is not evident for Fe, which had highly variable values among campaigns and locations. In 2007, Fe behaviour was similar to that found for Cl, Na, and SO_4 for locations 2, 3, and 4. The pH was not a differentiating parameter for the spatial and temporal contamination evolution, which can be explained by different production processes over time that generated either acid effluents (e.g., H_2SO_4 production) or alkaline effluents as a result of sodium and chlorate compounds production.

The most relevant aspect evidenced by these pictures is the huge increase in EC, Cl, Na, SO_4, and Fe values in 2007, in particular for points 2, 3, and 4. This is likely related to the construction of a sealed landfill in the period of 2003–2005, which served as a deposit of highly contaminated waste and soils. Operations associated with these works included mechanical relocation and remobilization of materials, likely promoting the oxidation and dissolution of contaminants hosted in those materials [25,45]. In 2016, this phenomenon was not observed and the values for these parameters were closer to those in the 1989 campaign. However, the areas corresponding to points 4 and 5 displayed higher values than those for points 2 and 3. This indicates that there was a dilution/attenuation of the contamination from 2007 to 2016, and, at same time, a migration of the plume to the SW, away from the ICCE.

5. Discussion

As demonstrated, geophysical methods provided an overall view of the area, but could not identify pollutants in groundwater or soils. On the other hand, hydrochemistry data could identify the pollutants present in the area. Before going into a more detailed discussion, it must be noted that the two methods were not used simultaneously and, hence, cannot be compared for the same fieldwork timeframe. Therefore, the discussion will be focused on temporal (multiannual) trends. It must also be noted, that, over the period under investigation, chemical plants changed production operations, while others closed down or started operations. Furthermore, analysis of contamination by organic material and by farming activities was not included because temporal data of these contaminants were not available.

Bearing in mind these points, it is clear that, according to the results obtained by both methodologies, ICCE activities have had a strong influence on the groundwater quality and have contributed to its degradation, as shown by both geophysical maps (Figures 4 and 6; Table 1) and hydrochemical data (Figures 5 and 9; Table 2).

Data from the pre-remediation period (before 2003–2005) showed lower levels of groundwater contamination, as revealed by smaller areas of higher conductivity in Figure 4, and some regression in contamination in some flow paths, but also a steady expansion of contamination in other paths (Table 1).

From 2001 to 2007, groundwater contamination peaked, as demonstrated by the larger high conductivity areas in Figure 4 and the increase in contamination velocity in Table 1, which was mainly due to higher concentrations of Cl, SO_4, Na, and Fe (Figure 9 and Table 2). During this period, the increase in concentrations may be explained by the remobilization of highly contaminated waste and soils when the sealed landfill was constructed and then filled. After this period, data from Figures 4–6 and Figure 9, as well as Tables 1 and 2, show that contamination attenuated over time and some shifting in space occurred with groundwater flow. In addition, some hotspots showed a spatial distribution (Figure 5) linked to the old, but still operating drainage effluent systems (ditches and underground pipes) used by the ICCE.

In conclusion, geophysics delivered an overall qualitative picture of the region, while hydrochemistry allowed for the qualification of the contaminant species. Furthermore, geophysical data provided information on the contamination spread velocity and paths, as well as on the effectiveness of remediation works and the impacts caused by wastes remobilization. Both methods allowed for the division of contamination characteristics in three periods: pre-remediation, the critical period while remediation works were implemented, and the post-remediation period. Some hotspots were identified, even in the 2016 campaign, pointing to new contamination sources associated with old effluent drainage systems that must be monitored and possibly repaired to stem the migration of contaminants.

Therefore, this combined methodology proved suitable in order to provide key information to stakeholders, technicians, and authorities on the evaluation and remediation operations carried out and those yet to be implemented. Finally, even if the methods are not applied simultaneously, it is clear that the methodology proved to be effective, fast, and reliable to predetermine location sampling points, and, thus, has a general applicability that extends beyond this case study.

Author Contributions: Conceptualisation: M.S.M., C.P. and E.F.d.S.; geochemical sampling campaigns: C.P., E.F.d.S., T.M. and N.D.; geophysical campaigns: M.S.M. and T.M.; hydrochemical analysis: C.P., E.F.d.S. and N.D.; geophysical data processing: M.S.M. and T.M.; writing—original draft preparation: C.P., M.S.M. and N.D.; writing—review and editing: M.S.M., N.D., C.P. and E.F.d.S. All authors have read and agreed to the published version of the manuscript.

Funding: This research was funded by the Portuguese Foundation for Science and the Technology (FCT), UID/GEO/04035/2020, and by Labex DRIIHM/Estarreja OHMI-International Observatory Hommes-Milieux (OHMI-CNRS), through the project "Combining geochemistry, geophysics and geostatistics to assist the spatial and temporal evolution of organic contamination in the aquifers of the Estarreja region: repercussions on biodiversity, agricultural practices, human and animal dietary habits".

Institutional Review Board Statement: Not applicable.

Informed Consent Statement: Not applicable.

Data Availability Statement: Geochemical and geophysical data presented in the paper can be found in the documents cited in the text with reference numbers [33,34,42,43].

Acknowledgments: N. Durães also thanks for his working support by national funds (OE), through FCT, in the scope of the framework contract foreseen in the numbers 4, 5 and 6 of the article 23, of the Decree-Law 57/2016, of August 29, changed by Law 57/2017, of July 19.

Conflicts of Interest: The authors declare no conflict of interest.

References

1. Gillham, R.W.; Sudicky, E.A.; Cherry, J.A.; Frind, E.O. An advection diffusion concept for solute transport in heterogeneous unconsolidated geological deposits. *Water Resour. Res.* **1984**, *20*, 369–378. [CrossRef]
2. Parker, B.L.; Chapman, S.W.; Guilbeault, M.A. Plume persistence caused by back diffusion from thin clay layers in a sand aquifer following TCE source-zone hydraulic isolation. *J. Contam. Hydrol.* **2008**, *102*, 86–104. [CrossRef] [PubMed]
3. Pilon-Smits, E. Phytoremediation. *Annu. Rev. Plant Biol.* **2005**, *56*, 15–39. [CrossRef]
4. Ibrahim, R.K.; Hayyan, M.; AlSaadi, M.A.; Hayyan, A.; Ibrahim, S. Environmental application of nanotechnology: Air, soil, and water. *Environ. Sci. Pollut. Res.* **2016**, *23*, 13754–13788. [CrossRef] [PubMed]
5. Blotevogel, J.; Mayeno, A.N.; Sale, T.C.; Borch, T. Prediction of contaminant persistence in aqueous phase: A quantum chemical approach. *Environ. Sci. Technol.* **2011**, *45*, 2236–2242. [CrossRef]
6. Khan, F.I.; Husain, T.; Hejazi, R. An overview and analysis of site remediation technologies. *J. Environ. Manag.* **2004**, *71*, 95–122. [CrossRef] [PubMed]
7. Vogelsang, D. *Environmental Geophysics, A Practical Guide*; Springer: Berlin, Germany, 1995; 173p.
8. Reynolds, J.M. *An Introduction to Applied and Environmental Geophysics*; Wiley Blackwell: Chichester, UK, 2011; 710p.
9. Greenhouse, J.P.; Monier-Williams, M.E.; Ellert, N.; Slaine, D.D. Geophysical methods in groundwater contamination studies. *Proc. Explor.* **1989**, *87*, 666–677.
10. BurVal Working Group. *Groundwater Resources in Buried Valleys—A Challenge for Geosciences*; Leibniz Institute for Applied Geosciences: Hannover, Germany, 2006; 303p.
11. Barker, R.; Moore, J. The application of time-lapse electrical tomography in groundwater studies. *Lead. Edge* **1998**, *17*, 1454–1458. [CrossRef]

12. Loke, M.H. Time-lapse resistivity imaging inversion. In Proceedings of the 5th EEGS-ES Meeting, Budapest, Hungary, 6–9 September 1999. [CrossRef]
13. Kim, J.-H.; Yi, M.-J.; Park, S.-G.; Kim, J.G. 4-D inversion of DC resistivity monitoring data acquired over a dynamically changing earth model. *J. Appl. Geophys.* **2009**, *68*, 522–532. [CrossRef]
14. Farzamian, M.; Monteiro Santos, F.A.; Khalil, M.A. Estimation of unsaturated hydraulic parameters in sandstone using electrical resistivity tomography under a water injection test. *J. Appl. Geophys.* **2015**, *121*, 71–83. [CrossRef]
15. Ulusoy, I.; Dahlin, T.; Bergman, B. Time-lapse electrical resistivity tomography of a water infiltration test on Johannishus Esker, Sweden. *Hydrogeol. J.* **2015**, *23*, 551–566. [CrossRef]
16. Cheng, P.-Y.; Chang, L.-C.; Hsu, S.-Y.; Tsai, J.-P.; Chen, W.-F. Estimating the hydrogeological parameters of an unconfined aquifer with the time-lapse resistivity-imaging method during pumping tests: Case studies at the Pengtsuo and Dajou sites, Taiwan. *J. Appl. Geophys.* **2017**, *144*, 134–143. [CrossRef]
17. Jouen, T.; Clément, R.; Henine, H.; Chaumont, C.; Vicent, B.; Tournebize, J. Evaluation and localization of an artificial drainage network by 3D time-lapse electrical resistivity tomography. *Environ. Sci. Pollut. Res.* **2018**, *25*, 23502–23514. [CrossRef]
18. Friedman, S.P. Soil properties influencing apparent electrical conductivity: A review. *Comput. Electron. Agric.* **2005**, *46*, 45–70. [CrossRef]
19. Appelo, C.A.J.; Postma, D. *Geochemistry, Groundwater and Pollution*, 2nd ed.; CRC Press: London, UK, 2005; 683p. [CrossRef]
20. Eissa, M.A.; Mahmoud, H.H.; Shouakar-Stash, O.; El-Shiekh, A.; Parker, B. Geophysical and geochemical studies to delineate seawater intrusion in Bagoush area, Northwestern coast, Egypt. *J. Afr. Earth Sci.* **2016**, *121*, 365–381. [CrossRef]
21. Kotra, K.K.; Yedluri, I.; Prasad, S.; Pasupureddi, S. Integrated geophysical and geochemical assessment for the comprehensive study of the groundwater. *Water Air Soil Pollut.* **2016**, *227*, 211. [CrossRef]
22. Najib, S.; Fadili, A.; Mehdi, K.; Riss, J.; Makan, A. Contribution of hydrochemical and geoelectrical approaches to investigate salinization process and seawater intrusion in the coastal aquifers of Chaouia, Morocco. *J. Contam. Hydrol.* **2017**, *198*, 24–36. [CrossRef]
23. Abbas, M.; Jardani, A.; Machour, N.; Dupont, J.-P. Geophysical and geochemical characterisation of a site impacted by hydrocarbon contamination undergoing biodegradation. *Near Surf. Geophys.* **2018**, *16*, 176–192. [CrossRef]
24. Brouwer, R.; Ordens, C.M.; Pinto, R.; Condesso de Melo, M.T. Economic valuation of groundwater protection using a groundwater quality ladder based on chemical threshold levels. *Ecol. Indic.* **2018**, *88*, 292–304. [CrossRef]
25. Costa, C.; Jesus-Rydin, C. Site investigation on heavy metals contaminated ground in Estarreja—Portugal. *Eng. Geol.* **2001**, *60*, 39–47. [CrossRef]
26. ERASE. *Estratégia de Redução dos Impactes Ambientais Associados aos Resíduos Industriais Depositados no Complexo Químico de Estarreja*; Environmental Impact Study; ERASE: Aveiro, Portugal, 2000.
27. ACAVACO. *Geophysical Prospecting in Estarreja*; Final Report; Process 1175; ACAVACO: Porto, Portugal, 1976. (In Portuguese)
28. Lucas, M.F.; Caldeira, M.T.; Hall, A.; Duarte, A.C.; Lima, C. Distribution of mercury in the sediments and fishes of the Lagoon of Aveiro, Portugal. *Water Sci. Technol.* **1986**, *18*, 141–148. [CrossRef]
29. Ferreira da Silva, E.A. *Impacte Ambiental de Elementos Maiores e Vestigiais no Aquífero Superficial de Estarreja: Caracterização da zona Envolvente ao Complexo de Estarreja*; Evidence of Scientific Aptitude and Pedagogical Proofs; University of Aveiro: Aveiro, Portugal, 1989.
30. Barradas, J.M.; Cardoso Fonseca, E.; Ferreira da Silva, E.; Garcia Pereira, H. Identification and mapping of pollution indices using a multivariate statistical methodology, Estarreja, central Portugal. *Appl. Geochem.* **1992**, *7*, 563–572. [CrossRef]
31. Matias, M.J.; Marques da Silva, M.A.; Carvalho, R.; Teles Roxo, G. *Estudo Geofísico Preliminar dos Aquíferos Quaternários a Oeste de Estarreja*; Atas da IV Conferência Nacional sobre a Qualidade do Ambiente: Lisboa, Portugal, 1994; pp. 184–198.
32. Neves, C. Estudo da contaminação das águas subterrâneas e respectivos processos de atenuação natural na zona industrial de Estarreja. Ph.D. Thesis, University of Lisbon, Lisbon, Portugal, 2015.
33. Ordens, C.M. Estudo da contaminação do aquífero superior na região de Estarreja. Master's Thesis, Faculdade de Ciências e Tecnologia, University of Coimbra, Coimbra, Portugal, 2007.
34. Marques, T.D.J. Estudo 4D da contaminação de aquíferos a oeste de Estarreja: Geoquímica e geofísica. Master's Thesis, University of Aveiro, Aveiro, Portugal, 2017.
35. Inácio, M.; Neves, O.; Pereira, V.; Silva, E.F. Levels of selected potential harmful elements (PHEs) in soils and vegetables used in diet of the population living in the surroundings of the Estarreja Chemical Complex (Portugal). *Appl. Geochem.* **2014**, *44*, 38–44. [CrossRef]
36. Cabral Pinto, M.A.S.; Marinho-Reis, A.P.; Almeida, A.; Ordens, C.M.; Silva, M.M.V.G.; Freitas, S.; Simões, M.; Moreira, P.I.; Dinis, P.A.; Diniz, M.L.; et al. Human predisposition to cognitive impairment and its relation with environmental exposure to potentially toxic elements. *Environ. Geochem. Health* **2018**, *40*, 1767–1784. [CrossRef] [PubMed]
37. Cabral Pinto, M.A.S.; Ordens, C.M.; Condesso de Melo, M.T.; Inácio, M.; Almeida, A.; Pinto, E.; Ferreira da Silva, E.A. An inter-disciplinary approach to evaluate human health risks due to long-term exposure to contaminated groundwater near a chemical complex. *Expo. Health* **2020**, *12*, 199–214. [CrossRef]
38. Teixeira, C.; Zbyszewski, G. *Carta Geológica de Portugal 1:50,000 & Notícia explicativa da Folha 16-A (Aveiro)*; Serviços Geológicos de Portugal: Lisboa, Portugal, 1976; 39p.

39. Condesso de Melo, M.T.; Marques da Silva, M.A. The Aveiro Quaternary and Cretaceous Aquifers, Portugal. In *Natural Groundwater Quality*; Edmunds, W.M., Shand, P., Eds.; Blackwell Publishing: Chennai, India, 2008; pp. 233–262.
40. Climate-Data.org. Available online: https://en.climate-data.org (accessed on 1 September 2020).
41. Butler, D.K. *Near-Surface Geophysics. Investigations in Geophysics*; No. 13; Society of Exploration Geophysicists: Tulsa, OK, USA, 2005.
42. Taunt, N. A Time-Lapse Electromagnetic Survey of a Contaminated Aquifer at Estarreja. Northern Portugal. Master's Thesis, University of Leeds, Leeds, UK, 2001.
43. Barradas, J.M. Geoquímica de Elementos Maiores e Vestigiais em Amostras de solos, Sedimentos de valas e Águas Subterrâneas. Contribuição para a Caracterização Ambiental da Zona Envolvente ao Complexo Químico de Estarreja. Ph.D. Thesis, University of Aveiro, Aveiro, Portugal, 1992.
44. McNeill, J.D. Use of electromagnetic methods for groundwater studies. In *Geotechnical and Environmental Geophysics*; Ward, S.H., Ed.; Society of Exploration Geophysicists: Tulsa, India, 1990; Volume I, pp. 191–218. [CrossRef]
45. Inácio, M.M.; Pereira, V.; Pinto, M.S. Mercury contamination in sandy soils surrounding an industrial emission source (Estarreja, Portugal). *Geoderma* **1998**, *85*, 325–339. [CrossRef]

applied sciences

MDPI

Article

Usefulness of Compiled Geophysical Prospecting Surveys in Groundwater Research in the Metropolitan District of Quito in Northern Ecuador

Lilia Peñafiel [1,2], Francisco Javier Alcalá [3,4,*] and Javier Senent-Aparicio [1]

1 Departamento de Ingeniería Civil, Universidad Católica de Murcia, 30107 Murcia, Spain; lapenafiel@alu.ucam.edu (L.P.); jsenent@ucam.edu (J.S.-A.)
2 Empresa Pública Metropolitana de Agua Potable y Saneamiento de Quito (EPMAPS), Quito 17-03-0330, Ecuador
3 Departamento de Desertificación y Geo-Ecología, Estación Experimental de Zonas Áridas (EEZA-CSIC), 04120 Almería, Spain
4 Instituto de Ciencias Químicas Aplicadas, Facultad de Ingeniería, Universidad Autónoma de Chile, Santiago 7500138, Chile
* Correspondence: fjalcala@eeza.csic.es; Tel.: +34-950-281-045

Abstract: As in other large Andean cities, the population in the Metropolitan District of Quito (MDQ) in northern Ecuador is growing, and groundwater is becoming essential to meet the increasing urban water demand. Quito's Public Water Supply Company (EPMAPS) is promoting groundwater research for sustainable water supply, and geophysical prospecting surveys are used to define aquifer geometry and certain transient groundwater features. This paper examines the usefulness of existing geophysical prospecting surveys in groundwater research in the MDQ. A database was built using 23 representative geophysical prospecting surveys compiled from EPMAPS' public repository, official geotechnical research reports, and the scientific literature. Fifteen EPMAPS-promoted surveys used near-surface electrical techniques (seven used electrical resistivity tomography and eight used vertical electrical sounding) to explore Holocene and Pleistocene sedimentary and volcano-sedimentary formations in the 25–500-m prospecting depth range, some of which form shallow aquifers used for water supply. Four other surveys used near-surface seismic techniques (refraction microtremor) for geotechnical research in civil works. These surveys have been reinterpreted to define shallow aquifer geometry. Finally, four surveys compiled from the scientific literature used electromagnetic techniques (magnetotelluric sounding and other very low-frequency methods) to explore Holocene to late Pliocene formations, some of which form thick regional aquifers catalogued as the larger freshwater reservoirs in the MDQ. However, no geophysical prospecting surveys exploring the complete saturated thickness of the Pliocene aquifers could be compiled. Geophysical prospecting surveys with greater penetration depth are proposed to bridge this research gap, which prevents the accurate assessment of the renewable groundwater fraction of the regional aquifers in the MDQ that can be exploited sustainably.

Keywords: geophysical prospecting techniques; groundwater research; urban water supply; Metropolitan District of Quito; Ecuador

Citation: Peñafiel, L.; Alcalá, F.J.; Senent-Aparicio, J. Usefulness of Compiled Geophysical Prospecting Surveys in Groundwater Research in the Metropolitan District of Quito in Northern Ecuador. *Appl. Sci.* **2021**, *11*, 11144. https://doi.org/10.3390/app112311144

Academic Editor: Hyung-Sup Jung

Received: 30 October 2021
Accepted: 22 November 2021
Published: 24 November 2021

1. Introduction

The Andean Highlands roughly extend between latitudes 11° N and 8° S, are over 3000 m a.s.l., play an important role in regional freshwater supply, and are highly sensitive to climate change [1–3]. As in other high mountain areas, most ecosystem typologies are groundwater dependent [2,4–6]. The combined influence of global driving forces and some anthropogenic activities (e.g., deforestation, overgrazing, soil degradation, and water overdevelopment) is altering river flow and aquifer recharge regimes [4–8], with negative consequences for human water supply and the preservation of dependent ecosystems [2,5,8]. Rivers and streams have traditionally been the main freshwater source to meet the water demand of downstream urban areas [4,9,10]. Increased demand in many densely populated Andean cities has driven water source diversification [2,4,6,10]. This is the case in the Metropolitan District of Quito city (MDQ) in northern Ecuador, where groundwater from the Andean Highlands and surface water transferred from the Amazonian watershed supplement traditional surface water sources [8,10]. A question arises of how is increasing groundwater extraction affecting reserves and dependent ecosystems? Applied groundwater research aimed at defining the aquifer functioning is yet incipient to answer this question accurately [11,12].

The Ecuador Water Authority and Quito's Public Water Supply Company (EPMAPS) are not immune to this problem. EPMAPS is responsible for prospecting, developing, distributing, and managing potable water in the MDQ, and promotes groundwater research to improve general hydrogeological knowledge as a prerequisite for a sustainable water supply. Hydrogeological studies use geophysical prospecting surveys to define aquifer geometry and certain transient groundwater features. Such studies typically cover two observation scales associated with two aquifer typologies. EPMAPS has used near-surface electrical geophysical techniques, such as electrical resistivity tomography (ERT) and vertical electrical sounding (VES), to explore shallow aquifer geometry and transient groundwater features required for drilling pumping wells intended to supply scattered population nuclei. The water company has also supported other public agencies and academic institutions can apply electromagnetic geophysical techniques for near-surface (very low-frequency electromagnetic methods, VLF-EMs) and deep (low-frequency magnetotelluric sounding, MTS) explorations to deduce the structure of shallow and thick regional geological formations catalogued as the larger freshwater reservoirs in the MDQ. Near-surface seismic prospecting techniques, such as refraction microtremor (REMI), have also been used to explore shallow geotechnical features.

Such geophysical prospecting techniques have proven useful in groundwater research in different hydrogeological environments [13–18]. They are non-invasive, usually inexpensive to apply, and useful when geotechnical sounding data is sparse or unable to provide subsurface information required for detailed groundwater research over multiple observation scales [15–17]. However, most geophysical prospecting surveys of interest in groundwater research are unpublished. Therefore, these experiences must be compiled and may need to be reinterpreted for groundwater research purposes. Such information concerning the aquifer saturated thickness, piezometric level, and spatial distribution of pore-water conductivity is essential to assess the fraction of groundwater that can sustainably supplement the urban water demand in the MDQ.

This paper examines the feasibility of compiled geophysical prospecting surveys in groundwater research in the MDQ, providing findings of interest to improve the hydrogeological conceptualization and identifying research gaps that can be bridged in the near future. Twenty-three representative geophysical prospecting surveys were compiled from EPMAPS' public repository, official geotechnical research reports, and the scientific literature. The compiled information was arranged in a database for interpretation. This paper does not intend to introduce new formulations, produce new data, discuss well-known principles of applied geophysical prospecting techniques, or assess the quality of the interpretations derived from the compiled surveys. This paper is organized as follows. Section 2 briefly describes the study area. Section 3 explains the steps followed to build the database. Section 4 reports the overall findings of the database analysis and gives an example of each geophysical prospecting technique. Section 5 discusses the geophysical prospecting scope, including research findings and gaps. Section 6 presents the main conclusions.

2. Study Area

2.1. Location and Climate

The MDQ is located in northern Ecuador at 0°14′ N to 0°35′ S and 78°10′ W to 78°56′ W (Figure 1), covers a surface area of 4320 km^2, and includes three main geomorphological sectors (Figure 1a). The 40-km-wide northern elongated Inter-Andean Valley (IAV) has an elevation ranging from 2100 to 3500 m a.s.l. and lies between the Western Andean Cordillera (WAC) (peak elevation 4776 m a.s.l., at Guagua-Pichincha Volcano) and Eastern Andean Cordillera (EAC) (peak elevation 4873 m a.s.l., at Sincholagua Volcano) (Figure 1b). The Guayllamba River flows north through the IAV and is the main surface watercourse (Figure 1b).

The MDQ exhibits a neo-tropical high-mountain climate, determined by the El Niño-Southern Oscillation and the Humboldt Current, and a steep orography [8,19,20]. Consequently, it has a marked distribution of biozones and ecosystems at different elevations including tropical mountain rainforests in the lowlands, wet alpine meadows (locally named *páramo*) in mid-slope areas, dry and cold scrublands in the highlands, and permanent snow covers at volcanoes' peaks [4,6,21].

Precipitation (P) follows a decreasing gradient from east to west, controlled by incoming Atlantic cloud fronts and elevation [22,23] and exhibits a positive gradient from low-lying areas to around 3500 m a.s.l. and a negative gradient above that elevation [19,20]. Most P occurs in February–May. In contrast, the lowest amount is recorded in July–September [23]. Annual mean P is around 1100 mm, with a coefficient of variation of 0.21 measured over the period 2003–2019. Annual mean temperature (T) is around 7.5 °C, with daily minimums in June–September and maximums in February–April. The decreasing T gradient with elevation is around 0.6–0.7 °C per 100 m elevation [21,22]. Insolation increases from low-lying areas to summits due to cloudiness induced by the Foehn effect in valleys from incoming Atlantic cloud fronts [7]. Annual mean potential evapotranspiration is around 1000 mm.

Table 1. Compiled information regarding the permeability and effective porosity of representative geological formations in the MDQ.

Lithology	Age	MDQ Sector [1]	Permeability [2]		Effective Porosity [3]	Reference
			Magnitude	Type		
Metapelites	Paleozoic	EAC	10^{-4}–10^{-2} (nd)	fr,fi	nd	[25,26]
Andesites and basalts	Cretaceous	WAC	10^{-4}–10^{-2} (nd)	fr,fi	nd	[25]
Sandstones and siltstones	Cretaceous	WAC	10^{-2}–10^{-1} (nd)	fr,fi	nd	[25]
Andesitic lavas	early Pleistocene	IAV	10^{-2}–10^{-1} (0.04)	fr,fi	nd	[26]
Andesitic lavas	middle Pleistocene	IAV	10^{-2}–10^{-1} (0.04)	fr,fi	0.02–0.08	[26,33,34]
Pyroclastic flows	middle Pleistocene	IAV	0.13–0.86 (nd)	fr,fi	nd	[25,26]
Ash	middle Pleistocene	IAV	10^{-3}–10^{-1} (0.01)	fr,fi	nd	[25,26]
Ash	late Pleistocene	IAV	10^{-3}–10^{-1} (0.01)	fr,fi	<0.01	[26,33,34]
Fluvio-glacial deposits	late Pleistocene	IAV	0.05–10 (1.02)	fi,ip	0.01–0.03	[25,26]
Ash	Holocene	IAV	10^{-3}–10^{-1} (nd)	fr,fi	nd	[25,26]
Avalanche flows	Holocene	IAV	10^{-2}–10^{-1} (nd)	fr,fi	nd	[25,26]
Lahar	Holocene	IAV	10^{-3}–10^{-2} (0.01)	fr,fi	0.01–0.06	[26,33,34]
Alluvial	Holocene	IAV	0.05–0.18 (0.12)	ip	0.05–0.12	[25,26]
Glacier and moraines	Holocene	IAV	0.05–0.15 (0.09)	ip	0.05–0.15	[25,26]

[1] EAC—Eastern Andean Cordillera, WAC—Western Andean Cordillera, and IAV—Inter-Andean Valley. [2] Permeability in m d^{-1}; magnitude refers to theoretical ranges and experimental values after borehole surveying (in parenthesis); fr—fracturation, fi—fissuration, and ip—intergranular porosity. [3] Effective porosity as a fraction; magnitude refers to experimental values after borehole surveying. nd—no data.

2.3. *Urban Water Demand*

Quito has historically been supplied from local rivers and streams. Since the 1990s, internal migration has produced rapid population growth in the MDQ, leading to increased water demand and the need to diversify water sources [37]. The MDQ currently has around 2.7 million inhabitants. Groundwater from the highlands (the EAC sector) and surface water transferred from the Amazonian watershed supplement the traditional surface water sources [8,10]. Groundwater exploitation began in the 1960s when the first pumping wells were drilled to supply northern urban districts [37]. Since then, EPMAPS has drilled more than 120 pumping wells to supply the increasing water demand [26]. The water supply system currently covers about 99% of the inhabitants, making the MDQ one of the best-served urban areas in Latin America. Groundwater meets around 16% of the total urban water demand. This figure will undoubtedly increase due to the noticeable population increase projected for the period 2020–2040 [37].

3. Data Compilation

A data search was conducted to examine the feasibility of existing geophysical prospecting surveys in groundwater research in the MDQ. The rationale was to create a database to cover as many geological formations (preferably those catalogued as aquifers), research interests, and prospecting techniques as possible. The selection prioritized geophysical surveys that explored depths of at least 10 m and used external validation data, such as geotechnical soundings logs and/or additional prospecting techniques. The selection also considered those surveys developed or promoted by the EPMAPS in sites where it has (or intends to have) operative water catchments. Therefore, EPMAPS' public repository (information available on request), official geotechnical research reports, and the scientific literature were consulted. Finally, 23 representative surveys covering the abovementioned scopes and priorities were selected to build the database in Table 2. Most of the compiled surveys were performed in the IAV sector (Figure 1b).

Table 2. Database comprising the compiled 23 geophysical prospecting surveys of interest in groundwater research in the MDQ. The data are clustered according to the applied geophysical prospecting technique, explored aquifer typology, deduced transient groundwater features, and additional technical information.

ID	Coordinates		Elevation, m a.s.l.	Geophysical Technique [1]			Geological Environment [2]						Research Interest [3]			Additional Technical Information [4]						Reference
				T1	T2	T3	G1	G2	G3	G4	G5	G6	R1	R2	R3	Variable	AQF	AQT	Profiles	PL	PD	
1	78°32′ W	0°24′ S	3046			b	a,b	a,b,c	a,b,c	c	a		a,b	b	a	ER, Ω m	10–210	220–8010	1	1300	1800	[18]
2	78°22′ W	0°18′ S	2644	a			a,b	a,b		a,b,c	a		a	b	b	ER, Ω m	62–170		10	400–500	250	[38]
3	78°22′ W	0°09′ S	2379	b			b	a,b,c		a,b,c	a		a	b	b	ER, Ω m	80–150		7	504–855	70–160	[39]
4	78°21′ W	0°06′ S	2350			a	b	a,b,c		a,b,c	a		a	b	b	ER, Ω m	15–195	210–280	13	160–750	130	[39]
5	78°20′ W	0°09′ S	2484	b			b	a,b,c		a,b,c	a		a	b	a,b	ER, Ω m	30–75	210–250	2	880	137	[40]
6	78°24′ W	0°04′ N	2064	b			b	a,b,c		b,c	a		a,b	a	a,b	ER, Ω m	10–50		2	303–358	50	[40]
7	78°42′ W	0°03′ N	1823	b			a,b	a,b		b,c		a	a	a	a,b	ER, Ω m	17–198	210–315	1	715	120	[40]
8	78°33′ W	0°17′ S	2860	b			a,b	a,b,c	a,b				a	a	c	ER, Ω m	3–210	215–300	9	250	40	[41]
9	78°29′ W	0°00′ N	2736	a			a,b	a,b	b,c	c			a	b	a	ER, Ω m	20–40		9	600–1000	322	[42]
10	78°25′ W	0°21′ S	2690	a			a	a,b,c		a,b,c			a	a	a,b	ER, Ω m	17–28	215–345	9	600–1000	271	[42]
11	78°30′ W	0°06′ S	2722	a			a,b	a,b,c	a,b,c	a,b,c			b	b	a,b	ER, Ω m	30–98		11	400–1000	200–500	[43]
12	78°32′ W	0°16′ S	2849		a		a,b	a,b,c	a,b,c				a	a	c	VS, m s^{-1}	95–680	95–680	171	8600	40–55	[44]
13	78°31′ W	0°13′ S	2826		a		a,b	a,b,c	c				b	b	c	VS, m s^{-1}	135–1050	135–1050	15	3200	120	[44]
14	78°29′ W	0°12′ S	2777		a		a,b	a,b,c	a,b,c				a	b	c	VS, m s^{-1}	125–710	125–710	171	10,330	40–55	[44]
15	78°11′ W	0°25′ S	4184			b	a,b		c	c	a	b	b	b	d	ER, Ω m	30–215	230–3190	130	15,000	4000	[45]
16	78°22′ W	0°02′ S	2145	a			a			a,b,c	a		a	b	a	ER, Ω m	50–170	218–457	23	600–1000	230	[46]
17	78°18′ W	0°18′ S	3260	a			a	a,b	c	c	a		b	b	a	ER, Ω m	28–56		3	600–800	180	[47]
18	78°33′ W	0°29′ S	2835	a			a,b	a,b,c	a	a,b,c	a		b	b	b	ER, Ω m	45–150	255–400	16	1000	200	[48]
19	78°29′ W	0°06′ S	2693	b			a,b	a,b	a,b				a	a	a,b	ER, Ω m	40–100	220–320	3	715	60	[49]
20	78°22′ W	0°12′ S	2280	b			a			a,b,c			a	a	c	ER, Ω m	20–150		3	110	25	[50]
21	78°22′ W	0°12′ S	2400		a		a	a,b,c					a	a	c	VS, m s^{-1}		720–945	7	120	60	[50]
22	78°25′ W	0°35′ S	3743	a					a,b,c	c	a		b	a	a	ER, Ω m	56–203	225–850	20	1000	150	[51]
23	78°30′ W	0°12′ S	2800			b	a	a,b,c	b,c	c	a	a	b	b	a	ER, Ω m	22–230	235–27,100	3	17,000	1500	[52]

[1] T1—Near-surface electrical techniques: (a) VES and (b) ERT. T2—Near-surface seismic techniques: (a) REMI. T3—Electromagnetic techniques: (a) VLF-EM and (b) Low-frequency MTS. [2] G1—Holocene sedimentary and volcano-sedimentary formations: (a) anthropogenic fillings and soils, and (b) silty and sandy ash. G2—late Pleistocene sedimentary and volcano-sedimentary formations: (a) ash and pumice ash, (b) tuff and paleo-soils, and (c) mud flows. G3—early–middle Pleistocene sedimentary and volcano-sedimentary formations: (a) lacustrine deposits, paleo-soils, peats, tuff, and microbreccias; (b) ash, pyroclastic flows, and tuff; and (c) avalanche flows, volcanic breccias and lavas. G4—early Pleistocene sedimentary and volcano-sedimentary formations: (a) alluvial sand and conglomerates; (b) siltstones and tuff; and (c) lahar, andesitic lavas, and avalanche flows. G5—late Pliocene volcanic formations: (a) andesitic lavas. G6—Basement: (a) Cretaceous WAC and IAV basement, and (b) Paleozoic EAC basement. [3] R1—Aquifer geometry: (a) layer thickness, and (b) fissuring and fracturing. R2—Aquifer dynamics and functioning in natural regime: (a) first groundwater observation, (b) regional piezometric level, (c) high-conductivity areas, and (d) geothermal areas. R3—Other applications: (a) basic research, (b) water supply, (c) civil works, and (d) geothermal energy. [4] AQF—Range of the variables electrical resistivity (ER) and share-wave velocity (VS) in geological formations catalogued as aquifers. AQT—Range of the variables ER and VS in geological formations catalogued as aquitards. PL—Prospecting length of the profiles in m. PD—Prospecting depth of the profiles in m.

Survey information was catalogued according to: (1) the applied prospecting technique; (2) the explored aquifer typology; (3) deduced transient groundwater features; and (4) additional technical information, such as the magnitude of the geophysical variables in geological formations catalogued as aquifers and aquitards, and prospecting length and depth (Table 2). The compiled data were initially checked to ensure a suitable statistical sample of the range of field technical conditions allowed by each geophysical technique. In all surveys, the original interpretations were examined to (i) standardize the age, lithological description, and hydrogeological behavior of the geological formations; (ii) adapt the achieved transient groundwater features to the scope of this paper; and (iii) adjust the drawing style to the scientific editing requirements. In the REMI and low-frequency MTS surveys, the original geotechnical and geodynamic findings were reinterpreted for groundwater research purposes.

Section 4 examines and classifies the compiled 23 surveys into three main groups of techniques (electrical, seismic, and electromagnetic) used to explore two main aquifer typologies (shallow and regional) and deduce two main research interests (aquifer geometry and transient groundwater features). This section also provides a representative survey of each technique. This survey represents an average condition of (i) prospecting depth and length, (ii) prospected geological formations, and (iii) deduced hydrogeological features regarding the compiled surveys of each technique.

4. Results

4.1. Near-Surface Electrical Surveys

Near-surface electrical techniques take voltage measurements between two potential electrodes installed on the land surface once direct current is injected into two current electrodes. Such techniques allow the calculation of subsurface electrical resistivity (ER) [Ω m], reciprocal of subsurface electrical conductivity (EC). Penetration depth and resolution depend on subsurface EC, which is a function of transient pore-water EC and steady ground EC, the input voltage used, and the electrode spacing adopted [53–55].

Fifteen surveys used near-surface electrical techniques; of these, eight surveys used VES for 1D ER models, and seven used ERT for 2D ER models. The VES surveys were part of groundwater research technical reports [38,42,43,46–48,51], whereas the ERT surveys included groundwater research technical reports [40,41,49,50] and scientific documents [39] (Table 2). The VES surveys aimed to define the punctual thickness of Holocene and late Pleistocene shallow aquifers. The ERT surveys aimed to explore the geometry of shallow aquifers and transient groundwater features, such as first groundwater observation, regional piezometric level, and pore-water conductivity. Both the VES and ERT surveys used Schlumberger, Wenner, and dipole–dipole arrays as the typical electrode configuration. The prospecting length range was 400–1000 m for the VES and 110–880 m for the ERT surveys, and the prospecting depth range was 150–500 m for the VES and 25–160 m for the ERT surveys (Table 2). For the geological formations catalogued as aquifers, the ER range was 17–203 Ω m for VES and 3–150 Ω m for ERT. For the geological formations catalogued as aquitards, the ER range was 215–850 Ω m for VES and 210–320 Ω m for ERT.

The ERT survey labelled 19 in Figure 1b and Table 2 was selected (Figure 2). This survey was performed in January 2016 and included three NNE–SSW ERT profiles (Figure 2b) covering a total prospecting length of 715 m [49]. The survey was part of groundwater research promoted by EPMAPS to supply scattered areas in the northern district of Quito city. Research interests were the geometry of shallow geological formations (some forming aquifers) and transient groundwater features, such as first groundwater observation.

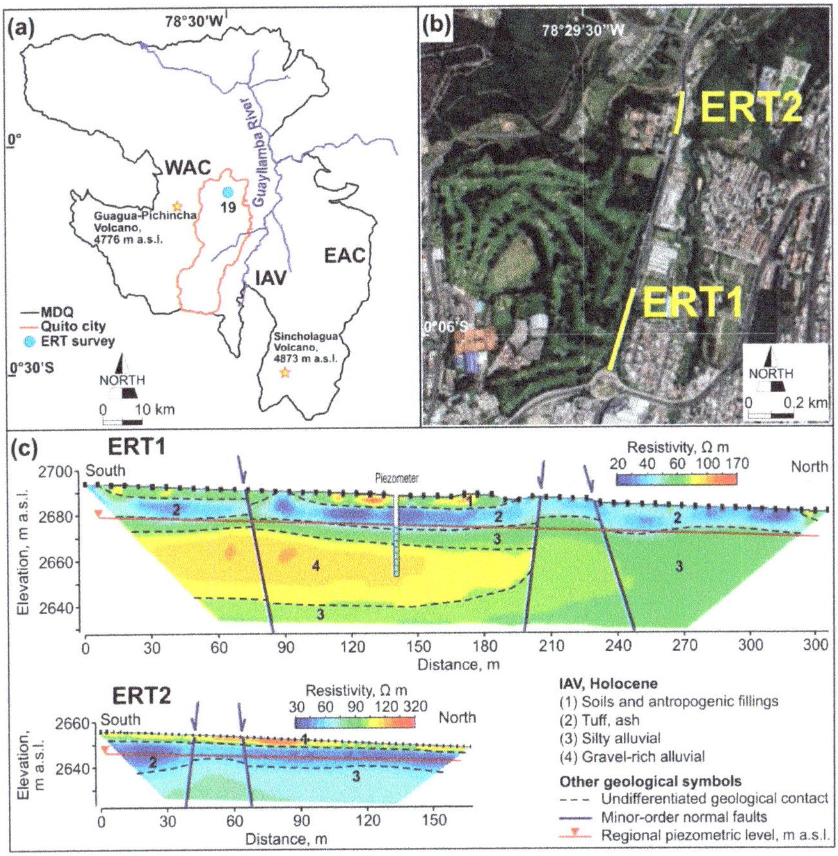

Figure 2. (**a**) General location of the selected ERT survey labelled 19 in Figure 1b and Table 2, updated and reinterpreted from Yautibug and Herrera [49]. (**b**) Detailed location of two selected ERT profiles, here called ERT1 and ERT2. (**c**) ERT1 and ERT2 profiles. The first groundwater observation (in this survey equivalent to the regional piezometric level) is singled out. Hydrogeological reinterpretation of ER models after [49] and local hydrogeological information [26,33,34]. Profiles are topographically corrected, and the vertical-to-horizontal scale ratio is 1:1.

The ER data were acquired using the SuperSting R8/IP eight-channels and the SuperSting R1/IP single-channel Memory Earth Resistivity and IP Meter by Advanced Geosciences Inc., Austin, TX. Fifty-six electrodes were placed along each ERT profile using a variable 3–6-m spacing and applying an input voltage of 200 V. A Schlumberger electrode array was used. See Yautibug and Herrera [49] for further methodological details.

Figure 2b shows the location of selected ERT profiles, here called ERT1 and ERT2. The profile features were prospecting lengths of 330 (ERT1) and 165 m (ERT2), prospecting depths of 59 (ERT1) and 28 m (ERT2), ER in the ranges of 20–175 (ERT1) and 25–330 Ω m (ERT2), and average root-mean-square errors (RMSE) of 1.80 (ERT1) and 2.94 (ERT2). Both ERT1 and ERT2 exposed quite similar horizontal and vertical ER distributions (Figure 2c). The ER values were typical of sedimentary and volcano-sedimentary rocks with a variable degree of saturation [56].

Local hydrogeological information [26,33,34] was used to reinterpret the ER models. From top to bottom, the vertical ER distribution was as follows: (i) 1–5 (ERT1) and 1–8 m (ERT2) of discontinuous porous soils and anthropogenic fillings with ER in the 70–300 Ω m range; (ii) 3–5 (ERT1) and 6–15 m (ERT2) of tuff and ash formation catalogued as aquitard with ER in the 20–50 Ω m range; (iii) 5–10 (ERT1) and 10–20 m (ERT2) of silty alluvial formation catalogued as aquitard with ER in the 40–80 Ω m range; (iv) 5–50 (ERT1) and 10–50 m (ERT2) of coarse-grained alluvial formation catalogued as aquifer with ER in the 80–120 Ω m range; and (v) 5–10 m (ERT1) of silty alluvial formation catalogued as aquitard with ER in the 40–80 Ω m range. The interbedded coarse-grained alluvial formation between the above silty alluvial formation (which includes items iii and v) is part of a shallow aquifer that provided the first groundwater observation corresponding to the regional piezometric level. Changes in the thickness and spatial continuity in the vertical ER distribution are due to sedimentary processes (e.g., lateral facies changes and erosive channels) and the action of minor-order normal faults described in the area [26,33,34].

4.2. Near-Surface Seismic Surveys

Near-surface seismic techniques respond to the steady shear modulus of subsurface materials, expressing seismic shear-wave velocity (VS) [L T^{-1}], in which the Rayleigh wave fundamental mode dispersion curve and higher modes (if present) are extracted from a shot record and then inverted to generate 1D VS models [57–62]. A succession of geophones records ambient microtremor to generate the Rayleigh waves from which a 2D VS model is obtained [61,62].

REMI was the near-surface seismic technique used to acquire VS data and map 2D VS models in four surveys designed to support geotechnical research in civil works (Table 2). The total prospecting length was 22.2 km, the prospecting depth was 40–120 m, and VS varied in the 95–1050 m s^{-1} range (Table 2).

This paper reinterpreted the 2D VS models for the shallow geological definition following the interpretative criteria reported by Paz et al. [63] and Alcalá et al. [64]. These authors propose that subsurface VS propagation is a site-specific steady property determined by effective compaction and therefore is dependent on the age and depth of each geological material piled vertically [65–68]. The different relationships between VS and age and depth in different lithologies described in the scientific literature [63–71] were used to reinterpret the VS models.

The VS increased in depth according to the increasing age and compaction of geological materials, from less than 200 m s^{-1} in recent anthropogenic fillings and lacustrine formations, 200–550 m s^{-1} in Holocene sedimentary and volcano-sedimentary formations, and more than 550 m s^{-1} in Pleistocene volcano-sedimentary formations. As in other near-surface seismic techniques, REMI cannot disambiguate boundaries of different geological formations with similar VS [63,64,70,71]. This limitation to make inner divisions was solved by using external validation data, such as regional [24,25] and local [26,33,34] geological information, geotechnical soundings logs, and other prospecting techniques [63,64].

The REMI surveys labelled 12 and 14 in Figure 1b and Table 2 were selected (Figure 3). They were performed in November 2011 as part of the Quito Subway geotechnical research, which included 201 REMI profiles grouped into three REMI surveys with prospecting lengths of 8.6 (southern Quito), 3.2 (central Quito), and 10.3 km (northern Quito) [44]. Since these prospecting lengths are too long to be drawn in detail, two 2.5-km sections from the southern- and northern-Quito REMI surveys exploring the most representative geological formations were selected (Figure 3b).

The VS data were acquired using the DAQlink-4 24-channels Compact Seismograph and the 4.5 Hz Geo-Space geophones by Seismic Source Co., Ponca city, OK, USA. The following configuration was applied: a recording array of 24 vertical component geophones, 4-m geophone spacing for a prospecting depth of around 40 m, 10-m displacement between readings, and a sampling rate of 0.25 m s^{-1}. See Cataldi [44] for further details about the data processing and mathematical inversion.

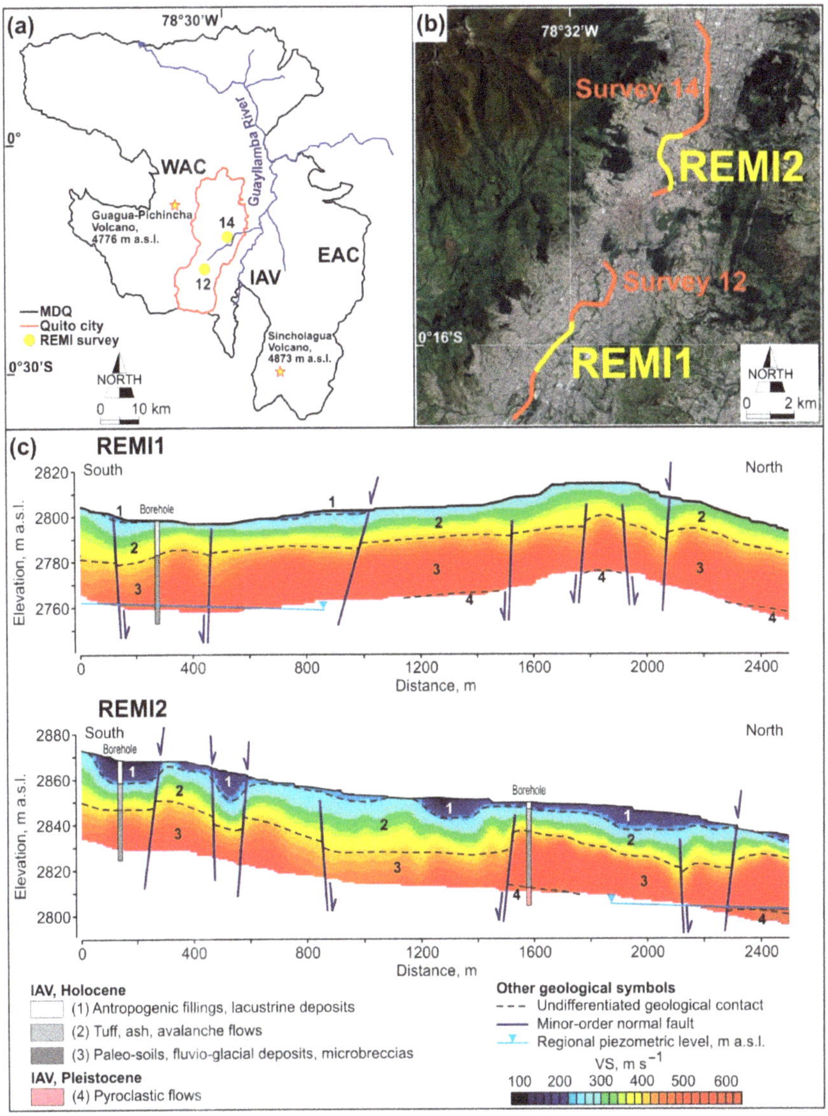

Figure 3. (**a**) General location of the selected southern- and northern-Quito REMI surveys labelled 12 and 14 in Figure 1b and Table 2, updated and reinterpreted from Cataldi [44]. (**b**) Detailed location of two selected 2.5-km sections from REMI surveys 12 and 14, here called REMI1 and REMI2. (**c**) REMI1 and REMI2 profiles' sections. The first groundwater observation (in these surveys equivalent to the regional piezometric level) is singled out. Hydrogeological reinterpretation of the two VS models after [44] and regional [24,25] and local [26,33,34] hydrogeological information. Profiles are topographically corrected, and the vertical-to-horizontal scale ratio is 1:0.13.

Figure 3b shows the location of the two selected REMI surveys sections, here called REMI1 from survey 12 and REMI2 from survey 14. The section features comprised an imposed prospecting length of 2.5 km, a prospecting depth of 40 m, VS in the range of 120–580 (REMI1) and 190–610 m s^{-1} (REMI2), and average RMSEs of 10.83 (REMI1) and 9.42 (REMI2) (Figure 3c). The VS values were typical of sedimentary [63,64] and volcano-sedimentary [72] rocks.

Based on geotechnical soundings logs and regional [24,25] and local [26,33,34] geological information, the vertical VS distribution was reinterpreted from top to bottom as follows: (i) 5–15 (REMI1) and 1–5 m (REMI2) of anthropogenic fillings, soils, and lacustrine deposits with VS less than 200 m s^{-1}; (ii) 10–40 (REMI1) and 10–20 m (REMI2) of Holocene tuff, ash, and avalanche flows with VS in the 200–400 m s^{-1} range; (iii) 1–40 m (REMI1) and less than 10 m (REMI2) of Holocene paleo-soils, fluvio-glacial deposits, and microbreccias with VS in the 400–550 m s^{-1} range; and (iv) 10–50 (REMI1) and 10–30 m (REMI2) of Pleistocene pyroclastic flows with VS higher than 550 m s^{-1}. The Pleistocene formations were only occasionally identified. The horizontal continuity of the vertical VS distribution was frequently interrupted by sedimentary processes (e.g., lateral facies changes and erosive channels) and the action of minor-order normal faults described in the urban area of Quito city [26,33,34].

4.3. Electromagnetic Surveys

Four surveys used electromagnetic techniques, specifically VLF-EM and low-frequency MTS, for the geometry and structure of Holocene to late Pliocene formations resulting from the action of first-order thrusts and strike-slip faults [18,39,45,52]. In general, electromagnetic techniques infer subsurface ER from measurements of natural geomagnetic and geoelectric field variations at the ground surface [73,74]. In particular, VLF-EM and low-frequency MTS use a fixed grounded dipole or horizontal loop as an artificial signal source to determine ER [75,76]. Both natural and controlled-source electromagnetic signals are used to obtain 1D ER models beneath the measurement site [77]. The ER dataset at the corresponding depths and signal-source distances are mathematically inverted to produce a 2D ER model [75,77].

One survey used VLF-EM (<10 Hz) in 13 profiles with a prospecting length in the 160–750 m range, a maximum prospecting depth of 130 m, and ER in the 15–280 Ω m range [39]. Three surveys used low-frequency MTS (>10 Hz) [18,45,52] with a total prospecting length of 33.3 km, a maximum prospecting depth of 4000 m, and ER in the ranges of 10–220 Ω m for geological formations catalogued as aquifers and 220–27,100 Ω m for geological formations catalogued as aquitards (Table 2).

The low-frequency MTS survey labelled 01 in Figure 1b and Table 2 was selected (Figure 4). This survey was performed in 2016 and included 13 measurement sites aligned in an NNE–SSW profile perpendicular to the Saguanchi Gorge strike-slip fault in the southern border of the MDQ (Figure 4b) [18]. Strike-slip faulting produces additional extensional areas disposed perpendicular to the primary shortening tectonic component evidenced by the Quito Fault System, which has implications for the drainage network and the extension and thickness of aquifers in the IAV [18].

The ER data were acquired using the StrataGem EH-4 four-channels Hybrid Source with a TxIM2 transmitter and electric BE-26 and magnetic G100k sensors by Geometrics, Inc., San Jose, CA, USA. The configuration was as follows: 13 measurement sites were set up using a variable 40–170 m array spacing and applying three frequencies in the 10–50 Hz range to record ER from depths of 0.6 to 1.8 km. The recorded ER data were mathematically inverted to obtain a 2D ER model. See Peñafiel et al. [18] for further methodological details.

Figure 4b shows the location of the selected low-frequency MTS profile, here called MTS1 (Figure 4c). The profile features included a prospecting length of 1300 m, a prospecting depth of 1800 m, ER in the 10–8010 Ω m range, and an average RMSE of 14.32. The ER values were similar to that reported for similar volcano-sedimentary rocks with variable degrees of fissuring, fracturing, and saturation [52,56].

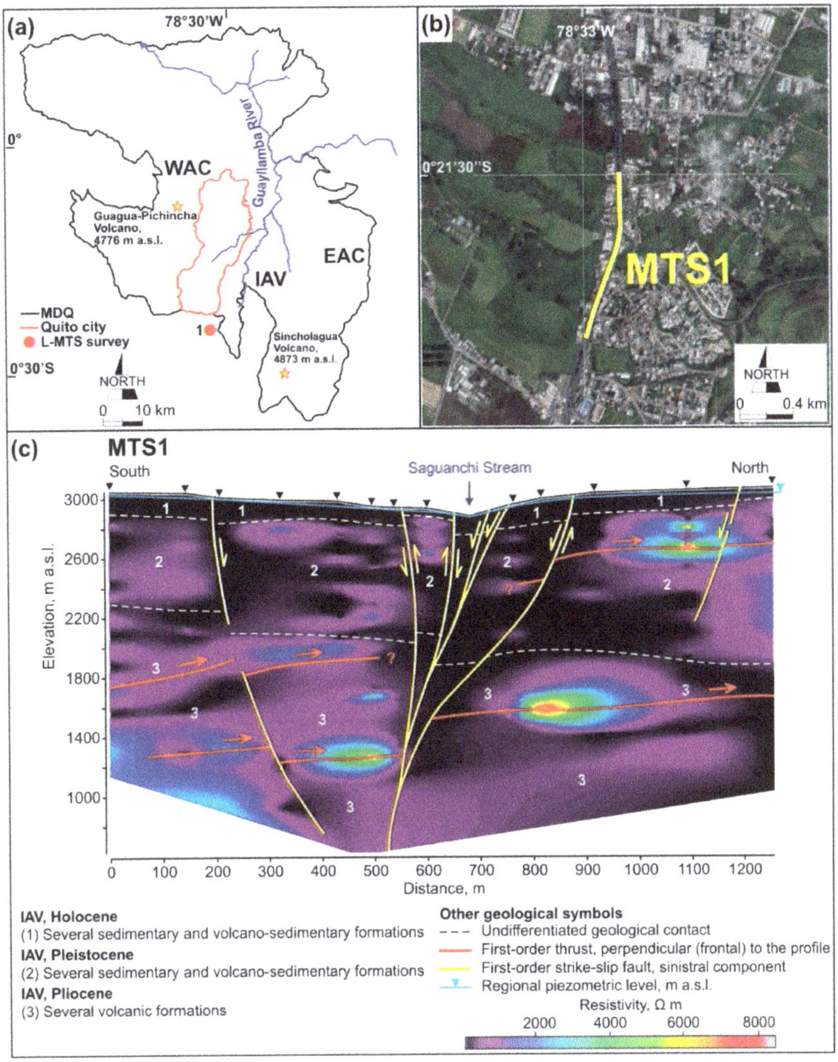

Figure 4. (**a**) General location of the selected low-frequency MTS survey labelled 1 in Figure 1b and Table 2, updated and reinterpreted from Peñafiel et al. [18]. (**b**) Detailed location of the selected low-frequency MTS profile, here called MTS1. (**c**) MTS1 profile. The regional piezometric level is singled out. Hydrogeological reinterpretation of the ER profile after [18] and regional [24,25] and local [26,33,34] hydrogeological information. Profile is topographically corrected, and the vertical-to-horizontal scale ratio is 0.34:1.

The ER model displayed (i) high-resistivity southern and northern sectors with an average ER around 1000 Ω m and several anomalies higher than 3000 Ω m, and (ii) a low-resistivity central sector bounded by strike-slip faults with an average ER lower than 1000 Ω m. In detail, ER values in the 10–50 Ω m range are associated with (i) strongly fractured volcanic rocks of different ages and high saturation degrees within the strike-slip fault zone, and (ii) sub-horizontal contacts within the Pleistocene and late Pliocene formations inferred by first-order thrusts observed at other sites (Figure 4c). ER values in the 50–200 Ω m range are associated with Holocene and Pleistocene volcano-sedimentary formations with moderate degrees of fissuring, fracturing, and saturation. ER values in the 200–1000 Ω m range are attributed to Pleistocene and late Pliocene volcanic formations with moderate fissuring and fracturing and low to moderate degrees of saturation. ER values higher than 1000 Ω m are interpreted as Pleistocene and late Pliocene volcanic rocks with low fissuring, fracturing, and saturation degrees.

5. Discussion

For the period 2020–2040, climate change projections foresee declining surface water sources in the Andean highlands, while the population in the MDQ could increase from 2.7 to about 4.2 million inhabitants [6,8,10]. The consequence is more groundwater exploitation to supplement the increasing urban water demand [37]. Water Authority of Ecuador and EPMAPS are aware of this problem and have already begun to promote applied groundwater research for sustainable use. Geophysical prospecting surveys can contribute to improve the hydrological conceptualization. However, the existing geophysical surveys explored different observations scales aimed to cover different research interests. So, the geophysical information must be examined and integrated before use in groundwater research. Most geophysical surveys have been performed in the most populated IAV sector (Figure 1b, Table 2), where groundwater exploitation is concentrated and signs of degradation have been reported [36,78].

To examine the usefulness of the subsurface geophysical information in groundwater research in the IAV, the area covered by each geophysical prospecting survey (defined by the corresponding prospecting length and depth) (Table 2) was superimposed onto a synthetic stratigraphic column in the southern border of the MDQ (Figure 1c). Of the compiled 23 geophysical prospecting surveys (Table 2), only the 20 ones located in the IAV (Figure 5) were selected and classified into three methodological groups (electrical, seismic, and electromagnetic) covering two observation scales and two aquifer typologies: shallow Holocene and late Pleistocene aquifers and thick regional middle–early Pleistocene and late Pliocene aquifers.

The VES and ERT surveys were used to define the geometry of shallow Holocene and late Pleistocene aquifers, deduce the regional piezometric level, and qualify pore-water conductivity. The prospecting depth was 150–500 m for the VES and 25–165 m for the ERT surveys (Figure 5). Reinterpretation of the 2D ER models shows that the ER range was 17–203 Ω m for VES and 3–150 Ω m for ERT in those geological formations catalogued as aquifers. These figures agree with the expected EC variability in saturated media associated with variable contributions of natural (e.g., recharge, thermalism, mineral dissolution) and anthropogenic (e.g., domestic, agriculture, industry) salinity sources. Groundwater conductivity deduced in shallow aquifers was higher than in thick regional aquifers. In those geological formations catalogued as aquitards, the ER range was 215–850 Ω m for VES and 210–320 Ω m for ERT. These figures agree with the expected lower variability of conductivity induced by the homogeneous clay content and barely variable lower pore-water content. The regional piezometric level varied depending on the aquifer hydraulic functioning, explored aquifer zone (recharge, transit, and discharge), and topography (Table 2).

REMI surveys were originally performed in geotechnical research for civil works. The 2D VS models were reinterpreted for the geometric definition of shallow Holocene and late Pleistocene aquifers, which is an innovative research application. The prospecting depth was 40 m (Figure 5). The VS values were less than 200 m s^{-1} in recent anthropogenic fillings, soils, and lacustrine formations; 200–550 m s^{-1} in Holocene formations; and more than 500 m s^{-1} in late Pleistocene formations.

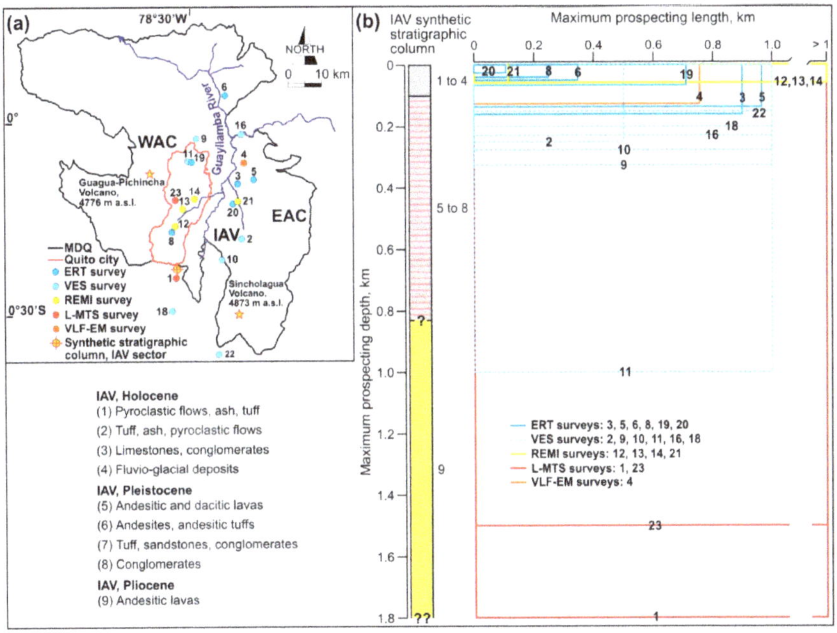

Figure 5. (**a**) General location of the 20 geophysical prospecting surveys in the IAV sector as shown in Figure 1b and Table 2; surveys 15 and 17 in the EAC sector and survey 7 in the WAC sector are excluded from this analysis. (**b**) A synthetic stratigraphic column of the IAV sector, as shown in Figure 1c. The typology and area covered by each geophysical prospecting survey are indicated. Acronyms ERT, VES, REMI, L-MTS, and VLF-EM are defined in Figure 1b.

VLF-EM and low-frequency MTS surveys provided the geometry and structure of Holocene to late Pliocene formations resulting from the action of first-order thrusts and strike-slip faults. The prospecting depths were 130 m for the VLF-EM survey and 1500, 1800, and 4000 m for the three low-frequency MTS surveys, of which only the first two were in the IAV sector (Figure 1b, Table 2). The hydrogeological reinterpretation of these two low-frequency MTS surveys [18,52] provided two significant findings: (i) the delineation of first-order thrusts and strike-slip faults controlling the geometry and stacking structure of Holocene to late Pliocene formations; and (ii) the identification of hitherto unknown disconnections (evidenced as high-resistivity fringes) between aquifers (evidenced as low-resistivity spaces) previously defined as hydraulically connected [26,36], resulting in less groundwater storage than previously known. An example is given in Figure 4c.

Despite the geophysical prospecting findings (Figure 5), three relevant gaps limiting a suitable hydrogeological conceptualization in the IAV sector still remain: (i) identifying the complete saturated thickness of Pliocene formations; (ii) elucidating the existence of older underlying Neogene formations of unknown hydrogeological behavior; and (iii) defining the IAV basement depth and typology, which is assumed to be equivalent to the WAC basement, after Bouguer gravity wedge data [31]. These gaps should be the subject of future research. Geophysical prospecting surveys with greater penetration depth could provide this basic aquifer information to assess the groundwater resource of Pleistocene and Pliocene andesitic lavas catalogued as the larger freshwater reservoirs in the MDQ.

6. Conclusions

The MDQ is a sparse-data area where definition of shallow and thick regional aquifers functioning, as well as their hydraulic relationships, is yet incipient. Different geophysical prospecting surveys originally devoted to different research interests can be integrated to provide subsurface information of interest in groundwater research. However, the published geophysical information is restricted to some research papers and scientific documents that aimed to investigate the transient groundwater features of shallow aquifers and the structure of deep (but not the deepest) geological formations. The compilation and examination of unpublished geophysical prospecting surveys contribute to improving the hydrogeological conceptualization, as well as to proposing additional research to bridge important gaps, which prevents the implementation of robust hydrological numerical tools to assess the groundwater resource.

A data search was conducted to examine the feasibility of existing geophysical prospecting surveys in groundwater research in the MDQ. Sources of information were the EPMAPS' public repository for near-surface electrical techniques (ERT and VES surveys), official geotechnical research reports in civil works for near-surface seismic techniques (REMI surveys), and scientific documents for electromagnetic techniques (MTS surveys). Finally, 23 representative geophysical prospecting surveys were compiled. Most of the surveys were performed in the IAV sector, where groundwater exploitation is concentrated. The ERT and VES surveys explored aquifer geometry and transient groundwater features of Holocene and late Pleistocene formations (some forming shallow aquifers), such as the aquifer saturated thickness, piezometric level, and spatial distribution of pore-water conductivity. The REMI surveys were reinterpreted to deduce the geometry of Holocene formations and, occasionally, late Pleistocene formations. The VLF-EM and low-frequency MTS surveys provided the structure of Holocene to late Pliocene formations in the IAV sector. No geophysical prospecting surveys exploring the complete saturated thickness of the Pliocene aquifers, other possible older underlying Neogene formations of unknown hydrogeological behavior, and the IAV basement depth and typology could be compiled. However, three surveys partially explored these features in the EAC and WAC sectors. Therefore, this basic information remains unknown, preventing an accurate assessment of the groundwater resource from which to deduce the renewable fraction of thick regional Pleistocene and Pliocene aquifers that can be exploited sustainably. Geophysical prospecting surveys with greater penetration depth could provide this basic information.

This paper demonstrates the need to systematize the use of geophysical prospecting techniques, including the most widely used technique described here to deduce shallow aquifer typologies and transient groundwater features and other specifics to explore the complete saturated thickness of Pleistocene and Pliocene aquifers forming the larger freshwater reservoirs in the MDQ. The above findings and research gaps, together with the generated database, seek to improve the design of geophysical prospecting surveys to explore groundwater resources in the MDQ and other large Andean urban areas.

Author Contributions: Conceptualization, F.J.A.; methodology, F.J.A.; formal analysis, L.P. and F.J.A.; data curation, L.P.; writing—original draft preparation, L.P., F.J.A. and J.S.-A.; writing—review and editing, F.J.A. and J.S.-A.; project administration, F.J.A.; funding acquisition, L.P. and F.J.A. All authors have read and agreed to the published version of the manuscript.

Funding: This research was partly supported by the International Atomic Energy Agency Project ECU/7/006 and the Ecuadorian Research Project PROMETEO-CEB-014-2015.

Institutional Review Board Statement: Not applicable.

Informed Consent Statement: Not applicable.

Data Availability Statement: The data presented in this study are available on request from the corresponding author.

Acknowledgments: The authors are grateful for the information on the geophysical prospecting surveys provided by the EPMAPS. We also wish to express our gratitude to two anonymous reviewers for their valuable advice and comments.

Conflicts of Interest: The authors declare no conflict of interest.

References

1. Barnett, T.P.; Adam, J.C.; Lettenmaier, D.P. Potential impacts of a warming climate on water availability in snow-dominated regions. *Nature* **2005**, *438*, 303–309. [CrossRef]
2. Bradley, R.S.; Vuille, M.; Diaz, H.F.; Vergara, W. Threats to water supplies in the Tropical Andes. *Science* **2006**, *312*, 1755–1756. [CrossRef] [PubMed]
3. Viviroli, D.; Archer, D.R.; Buytaert, W.; Fowler, H.J.; Greenwood, G.B.; Hamlet, A.F.; Huang, Y.; Koboltschning, G.; Litaor, M.L.; López-Moreno, J.L.; et al. Climate change and mountain water resources: Overview and recommendations for research, management and policy. *Hydrol. Earth Syst. Sci.* **2011**, *15*, 471–504. [CrossRef]
4. Buytaert, W.; Célleri, R.; De Bièvre, B.; Cisneros, F.; Wyseure, G.; Deckers, J.; Hofstede, R. Human impact on the hydrology of the Andean páramos. *Earth-Sci. Rev.* **2006**, *79*, 53–72. [CrossRef]
5. Buytaert, W.; Vuille, M.; Dewulf, A.; Urrutia, R.; Karmalkar, A.; Célleri, R. Uncertainties in climate change projections and regional downscaling in the tropical Andes: Implications for water resources management. *Hydrol. Earth Syst. Sci.* **2010**, *14*, 1247–1258. [CrossRef]
6. López, S.; Wright, C.; Costanza, P. Environmental change in the equatorial Andes: Linking climate, land use, and land cover transformations. *Remote Sens.* **2017**, *8*, 291–303. [CrossRef]
7. Flores-López, F.; Galaitsi, S.E.; Escobar, M.; Purkey, D. Modeling of Andean Páramo ecosystems hydrological response to environmental change. *Water* **2016**, *8*, 94–111. [CrossRef]
8. Gonzales-Zeas, D.; Erazo, B.; Lloret, P.; De Bièvre, D.; Steinchneider, S.; Dangles, O. Linking global climate change to local water availability: Limitations and prospects for a tropical mountain watershed. *Sci. Total Environ.* **2019**, *650*, 2577–2586. [CrossRef]
9. Alcalá, F.J.; Toapanta, J.; Peñafiel, L.; Barragán, E.; Yánez, W.; Buenaño, M.; Larrea, O. First data on atmospheric chloride mass balance components in the Andean páramo in central Ecuador: Implications to project climate scenarios of net aquifer recharge and potential groundwater chemical baseline. In *43rd IAH Congress: Groundwater and Society*; IAH: Montpellier, France, 2016; pp. 851–852.
10. Minga-Leon, S.; Gómez-Albores, M.A.; Bâ, K.M.; Balcázar, L.; Manzano-Solis, L.R.; Cuervo-Robayo, A.P.; Mastachi-Loza, C.A. Estimation of water yield in the hydrographic basin of southern Ecuador. *Hydrol. Earth Syst. Sci. Discuss.* **2018**, Preprint. [CrossRef]
11. Peñafiel, L.; Alcalá, F.J.; Barragán, E.; Larrea, O. Evaluación del balance hídrico en un área vulcano-sedimentaria de alta montaña poco monitorizada en la Cordillera de los Andes: Acuífero del río Pita, norte de Ecuador. *Rev. Lat. Hidrogeol.* **2016**, *10*, 502–508.
12. Peñafiel, L.; Alcalá, F.J.; Barragán, E.; Toro-Espitia, L.; Larrea, O. Uso de trazadores químicos e isotópicos para deducir el funcionamiento del acuífero del río Pita, Cordillera de los Andes, norte de Ecuador. *Rev. Lat. Hidrogeol.* **2016**, *10*, 495–501.
13. Paz, C.; Alcalá, F.J.; Carvalho, J.M.; Ribeiro, L. Current uses of ground penetrating radar in groundwater-dependent ecosystems research. *Sci. Total Environ.* **2017**, *595*, 868–885. [CrossRef]
14. Monteiro Santos, F.A.; Sultan, S.A.; Represas, P.; El Sorady, A.L. Joint inversion of gravity and geoelectric data for groundwater and structural investigation: Application to the northwestern part of Sinai, Egypt. *Geophys. J. Int.* **2006**, *165*, 705–718. [CrossRef]
15. Alam, K.; Ahmad, N. Determination of aquifer geometry through geophysical methods: A case study from Quetta Valley, Pakistan. *Acta Geophys.* **2014**, *62*, 142–163. [CrossRef]
16. Binley, A.; Hubbard, S.S.; Huisman, J.A.; Revil, A.; Robinson, D.A.; Singha, K.; Slater, L.D. The emergence of hydrogeophysics for improved understanding of subsurface processes over multiple scales. *Water Resour. Res.* **2015**, *51*, 3837–3866. [CrossRef] [PubMed]
17. Uhlemann, S.S.; Sorensen, J.P.R.; House, A.R.; Wilkinson, P.B.; Roberts, C.; Gooddy, D.C.; Binley, A.M.; Chambers, J.E. Integrated time-lapse geoelectrical imaging of wetland hydrological processes. *Water Resour. Res.* **2016**, *52*, 3. [CrossRef]
18. Peñafiel, L.; Reyes, P.S.B.; Alcalá, F.J.; Ramírez, M.R.; Cabero, A. Fold-axis parallel extension along the southern ending of the Quito (Ecuadorian Andes) fault system: Implications in river network and aquifer geometry. *Geotectonics* **2020**, *54*, 256–265. [CrossRef]
19. Bendix, J. Precipitation dynamics in Ecuador and northern Peru during the 1991/92 El Nino: A remote sensing perspective. *Int. J. Remote Sens.* **2000**, *21*, 533–548. [CrossRef]
20. Bendix, J.; Rollenbeck, R.; Palacios, W.E. Cloud detection in the Tropics–a suitable tool for climate-ecological studies in the high mountains of Ecuador. *Int. J. Remote Sens.* **2004**, *25*, 4521–4540. [CrossRef]
21. Van der Hammen, T.; Hooghiemstra, H. Neogene and Quaternary history of vegetation, climate, and plant diversity in Amazonia. *Quat. Sci. Rev.* **2000**, *19*, 725–742. [CrossRef]
22. Garreaud, R.; Falvey, M. The coastal winds off western subtropical South America in future climate scenarios. *Int. J. Climatol.* **2009**, *29*, 543–554. [CrossRef]
23. Rollenbeck, R.; Fabian, P.; Bendix, J. Precipitation dynamics and chemical properties in the mountain forests of Ecuador. *Adv. Geosci.* **2006**, *6*, 73–76. [CrossRef]
24. Burbano, N.; Becerra, S.; Pasquel, E. *Introduction to the Hydrogeology of Ecuador*; Memory and Appendices; National Institute for Meteorology and Hydrology: Quito, Ecuador, 2015; pp. 1–121.
25. PETROECUADOR. *Updated Mapping of Hydrogeological Units and Hydrographic Basins of Ecuador Referenced to the Hydrocarbon Sector Infrastructures (scale 1:1,000,000)*; Memory and Appendices; Government of Ecuador: Quito, Ecuador, 2005.

26. EPMAPS. *Hydrogeological Map of the Metropolitan District of Quito (scale 1:250,000)*; International Association of Hydrogeologists-Ecuadorian Group: Quito, Ecuador, 2012; Volume 2, pp. 25–30.
27. Trenkamp, R.; Kellogg, J.N.; Freymueller, J.T.; Mora, P. Wide plate margin deformation, southern Central America and northwestern South America, CASA GPS observations. *J. South Am. Earth Sci.* **2002**, *15*, 157–171. [CrossRef]
28. Tamay, J.; Galindo-Zaldívar, J.; Ruano, P.; Soto, J.; Lamas, F.; Azañón, J.M. New insight on the recent tectonic evolution and uplift of the southern Ecuadorian Andes from gravity and structural analysis of the Neogene-Quaternary intramontane basins. *J. S. Am. Earth Sci.* **2016**, *70*, 340–352. [CrossRef]
29. Hughes, R.A.; Pilatasig, L.F. Cretaceous and Tertiary terrane accretion in the Cordillera Occidental of the Ecuadorian Andes. *Tectonophysics* **2002**, *345*, 29–48. [CrossRef]
30. Litherland, M.; Aspden, J.; Jemielita, R.A. *The Metamorphic Belts of Ecuador. Nottingham*; British Geological Survey: Keyworth, UK, 1994; pp. 1–147.
31. Feininger, T.; Seguin, M.K. Simple Bouguer gravity anomaly field and the inferred crustal structure of continental Ecuador. *Geology* **1983**, *11*, 40–44. [CrossRef]
32. Villagómez, D. Evolución Geológica Plio-Cuaternaria del Valle Interandino Central en Ecuador (Zona de Quito–Guayllabamba–San Antonio). Master's Thesis, National Polytechnic School, Quito, Ecuador, 2003.
33. EPMAPS. *CAP-1 Borehole Drilling Report: The Pita Aquifer*; EPMAPS Repository: Quito, Ecuador, 2016.
34. EPMAPS. *CAP-2 Borehole Drilling Report: The Pita Aquifer*; EPMAPS Repository: Quito, Ecuador, 2016.
35. Muñoz, T. Contributions of Glacier Melting to the Upper Watershed of the Pita River, Ecuador. Master's Thesis, Michigan Technological University, Houghton, MI, USA, 2016.
36. Peñafiel, L. Geology and Analysis of the Groundwater Resource in the Southern Quito Subbasin. Master's Thesis, National Polytechnic School, Quito, Ecuador, 2009.
37. Larrea, O. *Los Acuíferos de Quito, Una Reserva Estratégica*; EPMAPS Repository: Quito, Ecuador, 2016.
38. Vaca, W.; Molano, M.; Heredia, G. *Estudio Geológico, Geofísico e Hidráulico en la Zona Industrial de Itulcachi*; Memory and Appendices; EPMAPS Repository: Quito, Ecuador, 2012.
39. Rios-Sanchez, M. A Remote Sensing Approach to Characterize the Hydrogeology of Mountainous Areas: Application to the Quito Aquifer System (QAS), Ecuador. Ph.D. Thesis, Michigan Technological University, Houghton, MI, USA, 2012.
40. Guachamín, A.; Vaca, W.; Naranjo, F.; Guzmán, G. *Elaboración de un Estudio de Tomografías Eléctricas en el Distrito Metropolitano de Quito (DMQ) y Análisis de Conductividades en Pozos*; Memory and Appendices; EPMAPS Repository: Quito, Ecuador, 2012.
41. Guerrero, M. *Informe Técnico Agencia Sur del Registro Civil*; Secretaría Nacional de Gestión de Riesgos: Quito, Ecuador, 2013; pp. 1–43.
42. Espinosa, V. *Investigaciones de Resistividad Eléctrica en la Exploración de Aguas Subterráneas: Pusuquí-San Antonio y Valle de Los Chillos*; Memory and Appendices; EPMAPS Repository: Quito, Ecuador, 2005.
43. Heredia, H. *Investigaciones Hidrogeológicas en el Acuífero de Quito, Sector El Condado*; Memory and Appendices; EPMAPS Repository: Quito, Ecuador, 2011.
44. Cataldi, A. *Estudio de Caracterización de Ruta con Métodos Geofísicos no Invasivos: Primera Línea del Metro de Quito*; Consorcio GRIFFMETAL–TRX Consulting: Quito, Ecuador, 2011; pp. 1–46.
45. Beate, B. *Estudio de Prefactibilidad Para Elaborar el Modelo Geotérmico Conceptual del Proyecto Chacana*; Memory and Appendices; Servicios y Remediación Serviremediación: Quito, Ecuador, 2012; pp. 1–183.
46. Espinosa, V. *Investigaciones Geofísicas de Resistividad Eléctrica en la Exploración de Aguas Subterráneas en el Proyecto El Quinche-Guayllabamba*; Memory and Appendices; EPMAPS Repository: Quito, Ecuador, 2007.
47. Hidalgo, J. *Proyecto de Agua Potable Ríos Orientales Portal de Salida del Túnel Papallacta–El Conde: Investigaciones Geofísicas*; Memory and Appendices; EPMAPS Repository: Quito, Ecuador, 2005.
48. Torres, A. *Actualización de Los Diseños del Proyecto de Agua Potable Tesalia*; Memory and Appendices; EPMAPS Repository: Quito, Ecuador, 2005.
49. Yautibug, G.; Herrera, F. *Investigaciones Mediante Métodos Geofísicos en la Avenida Córdova Galarza*; Memory and Appendices; EPMAPS Repository: Quito, Ecuador, 2016.
50. Vendramini, M.; Bermúdez, R.; Nionelli, P.; Erazo, M. *Diseño Definitivo Línea de Transmisión Paluguillo -Bellavista*; Memory and Appendices; EPMAPS Repository: Quito, Ecuador, 2017.
51. Beltrán, E. *Estudios de Prospección Geofísica Mediante Sondeos Eléctricos Verticales Realizados en la Cuenca Alta del Río Pita, Provincia de Pichincha*; Memory and Appendices; EPMAPS Repository: Quito, Ecuador, 2006.
52. Reyes, P.S.B.; Ramírez, M.R.; Cajas, M.I. Detecting a master thrust system by magnetotelluric sounding along the western Andean Piedmont of Quito, Ecuador. *Terra Nova* **2020**, *32*, 458–467. [CrossRef]
53. Hayley, K.; Bentley, L.R.; Gharibi, M.; Nightingale, M. Low temperature dependence of electrical resistivity: Implications for near surface geophysical monitoring. *Geophys. Res. Lett.* **2007**, *34*, L18402. [CrossRef]
54. Steelman, C.M.; Kennedy, C.S.; Capes, D.C.; Parker, B.L. Electrical resistivity dynamics beneath a fractured sedimentary bedrock riverbed in response to temperature and groundwater–surface water exchange. *Hydrol. Earth Syst. Sci.* **2017**, *21*, 3105–3123. [CrossRef]
55. Telford, W.M.; Geldart, L.P.; Sheriff, R.E. *Applied Geophysics*, 2nd ed.; Cambridge University Press: New York, NY, USA, 1990.

56. Ward, S.H. Resistivity and Induced Polarization Methods. In *Geotechnical and Environmental Geophysics*, 2nd ed.; Ward, S.H., Ed.; Society of Exploration Geophysicists: Tulsa, OK, USA, 1990; pp. 147–190.
57. Xia, J.; Miller, R.D.; Park, C.B. Estimation of near-surface shear-wave velocity by inversion of Rayleigh wave. *Geophysics* **1999**, *64*, 691–700. [CrossRef]
58. Xia, J.; Miller, R.D.; Park, C.B.; Hunter, J.A.; Harris, J.B.; Ivanov, J. Comparing shear-wave velocity profiles inverted from multichannel surface wave with borehole measurements. *Soil Dyn. Earthq. Eng.* **2002**, *22*, 181–190. [CrossRef]
59. Park, C.B.; Miller, R.D.; Xia, J. Multi-channel analysis of surface waves. *Geophysics* **1999**, *64*, 800–808. [CrossRef]
60. Park, C.B.; Miller, R.D.; Xia, J.; Ivanov, J. Multichannel analysis of surface waves (MASW)—Active and passive methods. *Lead Edge* **2007**, *26*, 60–64. [CrossRef]
61. Louie, J. Faster, better: Shear-wave velocity to 100 meters depth from refraction microtremor arrays. *Bull Seismol. Soc. Am.* **2001**, *91*, 347–364. [CrossRef]
62. Raines, M.G.; Gunn, D.A.; Morgan, D.J.R.; Williams, G.; Williams, J.D.O.; Caunt, S. Refraction microtremor (ReMi) to determine the shear-wave velocity structure of the near surface and its application to aid detection of a backfilled mineshaft. *Q. J. Eng. Geol. Hydrogeol.* **2011**, *44*, 211–220. [CrossRef]
63. Paz, M.C.; Alcalá, F.J.; Medeiros, A.; Martínez-Pagán, P.; Pérez-Cuevas, J.; Ribeiro, L. Integrated MASW and ERT imaging for geological definition of an unconfined alluvial aquifer sustaining a coastal groundwater-dependent ecosystem in southwest Portugal. *Appl. Sci.* **2020**, *10*, 5905. [CrossRef]
64. Alcalá, F.J.; Martínez-Pagán, P.; Paz, M.C.; Navarro, M.; Pérez-Cuevas, J.; Domingo, F. Combining of MASW and GPR Imaging and Hydrogeological Surveys for the Groundwater Resource Evaluation in a Coastal Urban Area in Southern Spain. *Appl. Sci.* **2021**, *11*, 3154. [CrossRef]
65. García-Jerez, A.; Navarro, M.; Alcalá, F.J.; Luzón, F.; Pérez-Ruiz, J.A.; Enomoto, T.; Vidal, F.; Ocaña, E. Shallow velocity structure using joint inversion of array and h/v spectral ratio of ambient noise: The case of Mula town (SE of Spain). *Soil Dyn. Earthq. Eng.* **2007**, *27*, 907–919. [CrossRef]
66. Mitchell, J.K.; Soga, K. *Fundamentals of Soil Behaviour*, 3rd ed.; Wiley: London, UK, 2005; pp. 1–592.
67. Zimmer, M.A.; Prasad, M.; Mavko, G.; Nur, A. Seismic velocities of unconsolidated sands: Part 1—Pressure trends from 0.1 to 20 MPa. *Geophysics* **2007**, *72*, E1–E13. [CrossRef]
68. McGann, C.R.; Bradley, B.A.; Cubrinovski, M. Investigation of shear wave velocity depth variability, site classification, and liquefaction vulnerability identification using a near-surface Vs model of Christchurch, New Zealand. *Soil Dyn. Earthq. Eng.* **2017**, *92*, 692–705. [CrossRef]
69. Alcalá, F.J.; Espinosa, J.; Navarro, M.; Sánchez, F.J. Propuesta de división geológica de la localidad de Adra (provincia de Almería). Aplicación a la zonación sísmica. *Rev. Soc. Geológica España* **2002**, *15*, 55–66.
70. Martínez-Pagán, P.; Navarro, M.; Pérez-Cuevas, J.; Alcalá, F.J.; García-Jerez, A.; Sandoval-Castaño, S. Shear-wave velocity based seismic microzonation of Lorca city (SE Spain) from MASW analysis. *Near Surf. Geophys.* **2014**, *12*, 739–749. [CrossRef]
71. Martínez-Pagán, P.; Navarro, M.; Pérez-Cuevas, J.; Alcalá, F.J.; García-Jerez, A.; Vidal, F. Shear-wave velocity structure from MASW and SPAC methods. The case of Adra town, SE Spain. *Near Surf. Geophys.* **2018**, *16*, 356–371. [CrossRef]
72. Vanorio, T.; Prasad, M.; Patella, D.; Nur, A. Ultrasonic velocity measurements in volcanic rocks: Correlation with microtexture. *Geophys. J. Int.* **2002**, *149*, 22–36. [CrossRef]
73. Unsworth, M.; Soyer, W.; Tuncer, V.; Wagner, A.; Barnes, D. Hydrogeologic assessment of the Amchitka Island nuclear test site (Alaska) with magnetotellurics. *Geophysics* **2007**, *72*, B47–B57. [CrossRef]
74. Chave, A.D.; Jones, A.G. *The Magnetotelluric Method: Theory and Practice*; Cambridge University Press: New York, NY, USA, 2012; pp. 1–584.
75. McPhee, D.K.; Chuchel, B.A.; Pellerin, L. *Audiomagnetotelluric Data from Spring, Cave, and Coyote Spring Valleys, Nevada*; No. 2006-1164; US Geological Survey: Menlo Park, CA, USA, 2006; pp. 1–43.
76. Zonge, K.L.; Hughes, L.J. Controlled source audio-frequency magnetotellurics. In *Electromagnetic Methods in Applied Geophysics, Volume 2: Application, Parts A and B*; Nabighian, M.N., Ed.; Society of Exploration Geophysics: McLean, VA, USA, 1991; pp. 713–810.
77. Geometrics, Inc. *Operation Manual for Stratagem Systems Running IMAGEM Ver. 2.19*; Geometrics, Inc.: San Jose, CA, USA, 2007.
78. Buytaert, W.; De Bièvre, B. Water for cities: The impacts of climate change and demographic growth in the tropical Andes. *Water Resour. Res.* **2012**, *48*, W08503. [CrossRef]

applied sciences

MDPI

Article

Geophysical Characterization in the Shallow Water Estuarine Lakes of the Southern Everglades, Florida

Michael Eyob Kiflai [1,2,*], Dean Whitman [2], René M. Price [2,3], Thomas A. Frankovich [3] and Christopher J. Madden [4]

1 Division of Water Rights, State Water Resources Control Board, Sacramento, CA 95814, USA
2 Department of Earth and Environment, Florida International University, Miami, FL 33199, USA; whitmand@fiu.edu (D.W.); pricer@fiu.edu (R.M.P.)
3 Institute of Environment, Florida International University, Miami, FL 33199, USA; tfrankov@fiu.edu
4 South Florida Water Management District, Everglades Systems Research Division, West Palm Beach, FL 33406, USA; cmadden@sfwmd.gov
* Correspondence: michael.kiflai@waterboards.ca.gov or mkifl001@fiu.edu

Abstract: Anthropogenic activities have greatly modified freshwater flows through Everglades National Park (ENP) such that saltwater has intruded extensively inland from the coastline, causing coastal lakes and their ecosystems to be exposed to varying salinity conditions. The Comprehensive Everglades Restoration Plan (CERP) makes an effort to restore the quantity, quality, timing, and distribution of freshwater flow in ENP with a goal of reducing salinity conditions within the coastal communities and adjacent estuaries. An understanding of the temporal and spatial variations of surface water and shallow groundwater salinity in the coastal lakes of ENP is needed to evaluate restoration efforts. Geophysical surveys were conducted between 2016 to 2019 using electrical resistivity and electromagnetic (EM) methods in the coastal lakes of ENP. A mean local formation factor of 10.7 ± 1.8 was calculated for the region by comparing the lakes' bottom formation inverted electrical resistivity soundings with coincident pore water resistivity measured in groundwater wells. The conductivity of surface and groundwater increased during the dry season, reflecting decreased precipitation, increased evapotranspiration, and the increasing influence of saline water from Florida Bay. Spatially, salinity in the lakes increased from west to east in the surface water with an opposite trend observed in the shallow groundwater. Along the south to north inland direction, the salinity of both surface water and groundwater decreased. This study demonstrates that floating electrical resistivity and EM methods can characterize the subsurface formation resistivity and describe temporal and spatial patterns of surface and shallow groundwater conductivity.

Keywords: Everglades National Park (ENP); electrical resistivity; electromagnetism; formation factor; salinity and constrained inversion

Citation: Kiflai, M.E.; Whitman, D.; Price, R.M.; Frankovich, T.A.; Madden, C.J. Geophysical Characterization in the Shallow Water Estuarine Lakes of the Southern Everglades, Florida. *Appl. Sci.* **2022**, *12*, 1154. https://doi.org/10.3390/app12031154

Academic Editors: Francisco Javier Alcalá, Maria Catarina Paz, Pedro Martínez-Pagán and Fernando Monteiro Santos

Received: 1 November 2021
Accepted: 13 January 2022
Published: 22 January 2022

1. Introduction

During the past century, Everglades National Park (ENP) has been adversely impacted by past human activities that have altered the flow of freshwater through the system [1]. Moreover, the underlying highly permeable limestone aquifer is susceptible to saltwater intrusion (SWI) along the coastline [2]. In the 1950s, the U.S. Army Corps of Engineers developed canals, levees, and water conservation areas for flood protection. Even though this development plan has succeeded in controlling floods, South Florida does not receive sufficient quantity and distribution of water, which results in ecosystem degradation and extensive saltwater intrusion [1].

In 2000, the U.S. Congress authorized the Comprehensive Everglades Restoration Plan (CERP) to restore, preserve, and protect the South Florida ecosystem. The CERP makes an effort to restore the quantity, quality, timing, and distribution of freshwater in the region [3]. Major activities of the CERP are expected to increase the flow of freshwater and modify the groundwater chemistry in the ENP. Freshwater input to the ENP comes from direct rainfall and inflows from the water conservation area (WCA) reservoirs to the north [4] (Figure 1). The two main flow drainage paths in ENP are the Shark River Slough (SRS) and Taylor Slough (TS) (Figure 1). The two lake regions studied in this project are located between these two main flow pathways. Salinity in Florida Bay varies in time and space and is governed by the influence of precipitation, evapotranspiration, runoff, and mass exchange with the surrounding basins [5]. Evapotranspiration rates are high in this region and affect the freshwater input balance negatively [4]. The author also suggests freshwater inflow is critical to compensate for the net loss.

The degree of saltwater intrusion along the coast varies widely and is affected by the hydroclimate variability, hydrogeological setting, history of groundwater development, and sources of saline water within a particular area [6]. Under natural conditions, the seaward flow of freshwater prevents saltwater from encroaching coastal aquifers. However, in comparison with the historical flow of freshwater, the flow of freshwater through the Everglades has been reduced by approximately 70%. Due to this, water quality has degraded, and nearshore dry-season salinities increased by 20 to 30 PSU [7]. This reduced historical flow of freshwater input to the Florida Bay followed by drought in South Florida between 2014 and 2015 produced a substantial die-off of the seagrass [8].

Most commonly, mapping of saltwater intrusion in coastal aquifers has been conducted by collecting water samples from wells. Even though this approach provides reliable results, it is restricted to the existing and accessible wells. Geophysical methods have been developed to identify and map the lateral and vertical distribution of shallow freshwater and saline water interface. These methods provide powerful tools to identify the position of saline or brackish water in an aquifer [9]. Electrical resistivity and electromagnetics are the most commonly used geophysical methods to study saltwater intrusion. Integrated approaches of electromagnetics (EM) and direct current resistivity were applied in the coastal wetland of Tampa Bay, Florida [10] and Azan Basin, Jordan [11]. Monsoor et al. [12] used EM 31 for rapid characterization of shallow water, the contaminated wetlands of Kearny Marsh, New Jersey. EM methods have been used for imaging saltwater intrusion in coastal aquifers—namely the seaside groundwater basin, California [13]—and for characterizing the chemical properties of soil in Sidrolandia, MS, Brazil [14].

Electrical resistivity has an advantage because the non-uniqueness of the resistivity method is less compared with other geophysical methods such as electromagnetic methods [15]. The electrical resistivity method is superior for imaging the electrical resistivity structure compared to other non-invasive geophysical imaging techniques. DC resistivity offers several advantages over EM methods because the array is in direct galvanic contact with the water or earth and is not affected by external magnetic fields and nearby conducting bodies such as the boat and operator. The EM method provides a faster alternative to DC soundings [16], but it is sensitive to electromagnetic noise produced by metals, pipelines, etc. Therefore, the DC resistivity sounding remains the preferred method, though an integrated approach using both methods is complementary.

The integrated approaches of EM, electrical resistivity, and well data are the most effective technique to estimate a formation factor, which is used to produce a regional map of groundwater salinity. A formation factor is the ratio of the resistivity (conductivity) of a rock filled with water to the resistivity (conductivity) of that water. It can be determined by coincident measurements of groundwater and formation conductivity or from empirical relationships such as Archie's Law [17]. An effective technique to map pore water conductivities is to compute a formation factor from resistivity surveys and pore water samples. An integrated approach of EM and electrical resistivity was conducted on the wetland of Tampa Bay, Florida [10] and computes a formation factor from resistivity surveys and pore water samples. The study conducted by [18] estimated a formation factor of 5.1 by comparing EM induction logs and water samples from wells in eastern Miami-Dade County. Similarly, in other studies, Ref. [19] estimated a formation factor of 9.65 in Everglades National Park using similar methods and [20] estimated a formation factor of 9.8 for Big Pine Key using ERT and water samples from wells. The formation factor depends on several parameters such as porosity, the degree of saturation, cementation, and pore fluid resistivity and heterogeneity.

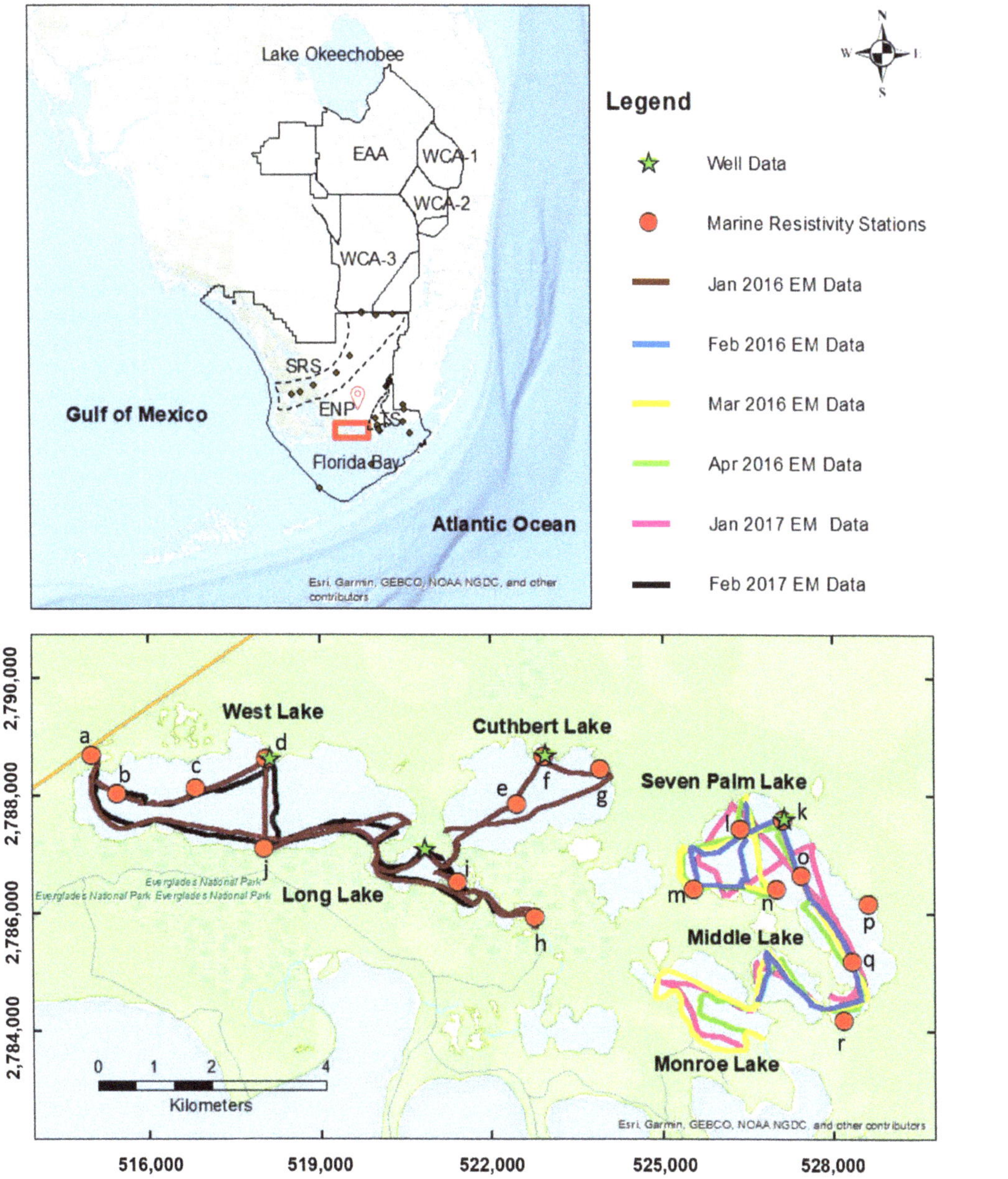

Figure 1. Map of South Florida showing the location of Everglades National Park (ENP), Everglades Agricultural Areas (EEA), Water Conservation Area (WCA), Shark River Slough (SRS), Taylor Slough (TS), Meteorologic NCL station (Red Pin), well stations, and geophysical survey. The SRS flows from the north to the southwest into the Gulf of Mexico, and the TS flows south into Florida Bay. The geophysical survey includes EM survey denoted by lines and floating array electrical resistivity denoted by letters from a to r. Map coordinates are in UTM, Zone 17N.

A number of geophysical studies have been conducted in the ENP. A study conducted in the southern ENP [19] aimed to assess saltwater intrusion and prepared a subsurface resistivity map using airborne electromagnetic methods (FDEM) and borehole geophysical measurements. The authors established correlations between formation resistivity and water-specific conductance for the region and provided a baseline for further studies. In 2011, a TDEM survey was conducted in Miami-Dade County and delineated the location of the freshwater/saltwater interface, as well as showing the influence of canals and roadbeds on the hydrologic regime [19,21]. Surveys of surface water salinities of the coastal lakes within ENP indicated brackish to hypersaline salinities that varied according to season (higher in the summer following the dry season), and proximity of upstream freshwater sources and exchange with Florida Bay marine waters downstream [22,23].

This paper aims to present the relationship between the freshwater input and hydrochemistry of surface waters and lake-bottom groundwater in shallow (1–2 m) brackish lakes within the southern Everglades known as the Mangrove Lakes (Figure 1). It explores the spatial and temporal changes in groundwater chemistry using electrical resistivity, EM methods, surface water, and groundwater sampling. The EM surveys were conducted from 2016 to 2017 using a GSSI EMP-400 Profiler multi-frequency [24] EM conductivity meter integrated with a GPS receiver deployed in a flat-bottomed plastic kayak towed behind a motorized jon boat. In addition, electrical resistivity soundings were conducted at spot locations in July 2019 using a floating electrode array connected to an Advanced Geosciences, Inc. (AGI) Super Sting resistivity meter [25]. During the survey, at various spot locations, the surface water conductivity, temperature, pH, and salinity were recorded using a YSI water quality data sonde. Water depths were also measured. Groundwater-specific conductivity was continuously monitored at four shallow wells on the shorelines of the lakes and was compared to nearby inverted lake bottom resistivity to calculate a formation factor for the Lakes region. This formation factor was used to convert the inverted formation resistivity to groundwater resistivity and produce a regional map of groundwater salinity. This study investigated the spatial and temporal changes in the surface water and groundwater salinity in the Mangrove Lakes of ENP using electrical resistivity and EM methods. Understanding the hydrochemistry of the aquifer can help to establish sustainable water resources management, and regularly monitoring the hydrological conditions is required for proper water management practice and conservation actions. This study showed that floating electrical resistivity arrays and EM ground conductivity meters can effectively characterize the lake bottom salinity and can be used regularly in monitoring the surface water and groundwater salinity in shallow water bodies.

2. Electrical and Electromagnetic Methods

Geophysical methods such as electrical resistivity and electromagnetics are rapid and noninvasive geophysical methods for measuring groundwater properties and characterizing the spatial and temporal variability of subsurface formations [26]. These methods include electrical resistivity methods such as vertical electrical soundings (VES) and electrical resistivity tomography (ERT), and electromagnetic (EM) methods such as frequency-domain electromagnetic (FDEM) and time-domain electromagnetic (TDEM) soundings and profiling. These methods are commonly used in near-surface geophysics because the subsurface electrical properties are easily correlated to the physical and chemical properties of fluids within the pore space [27]. These methods can produce a high-resolution image of the shallow subsurface formation and have been widely used in hydrogeology and environmental studies.

In electrical resistivity, a current is injected across a pair of electrodes, and the voltage difference between the potential electrodes is measured. The voltage difference is a function of the injected current and the resistivity beneath the electrode array. In the electrical resistivity method, commonly used electrode configurations include the Wenner, Schlumberger, and dipole–dipole arrays. In dipole–dipole arrays, the dipoles are equal in width (a) and separated by a distance $n \times a$, where n is an integer multiplier (Figure 2B).

Electromagnetism uses the principles of electromagnetic induction to map variations in electrical conductivity. EM waves from a transmitting coil induce eddy currents in subsurface conductors, resulting in a secondary magnetic field detected in a receiver coil. The electrical resistivity and EM data are modeled to infer the electrical conductivity of the subsurface.

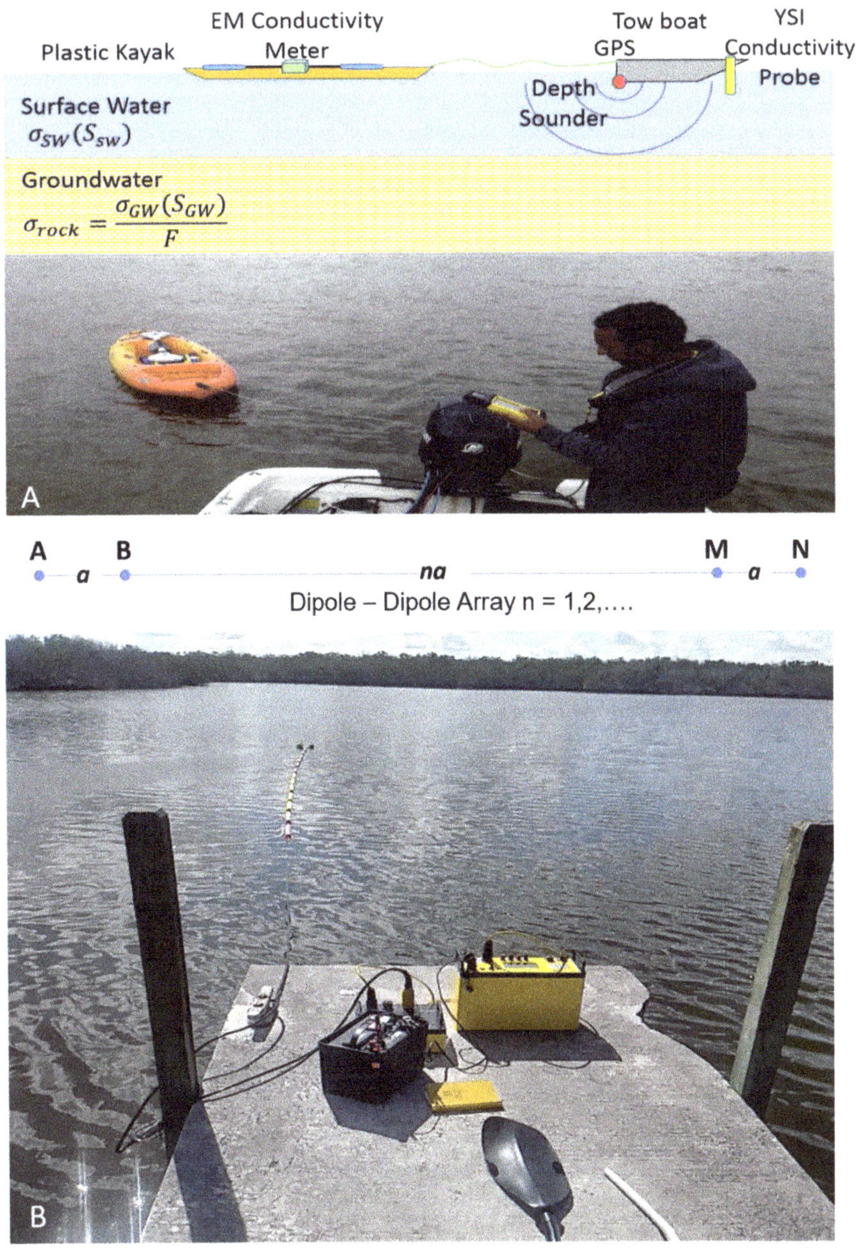

Figure 2. Geophysical experimental set up in West Lake, Everglades National Park, FL. (**A**) EM experimental set up. (**B**) Electrical resistivity experimental set up. In the dipole–dipole array, A,B are the current electrodes. M and N are the potential electrodes.

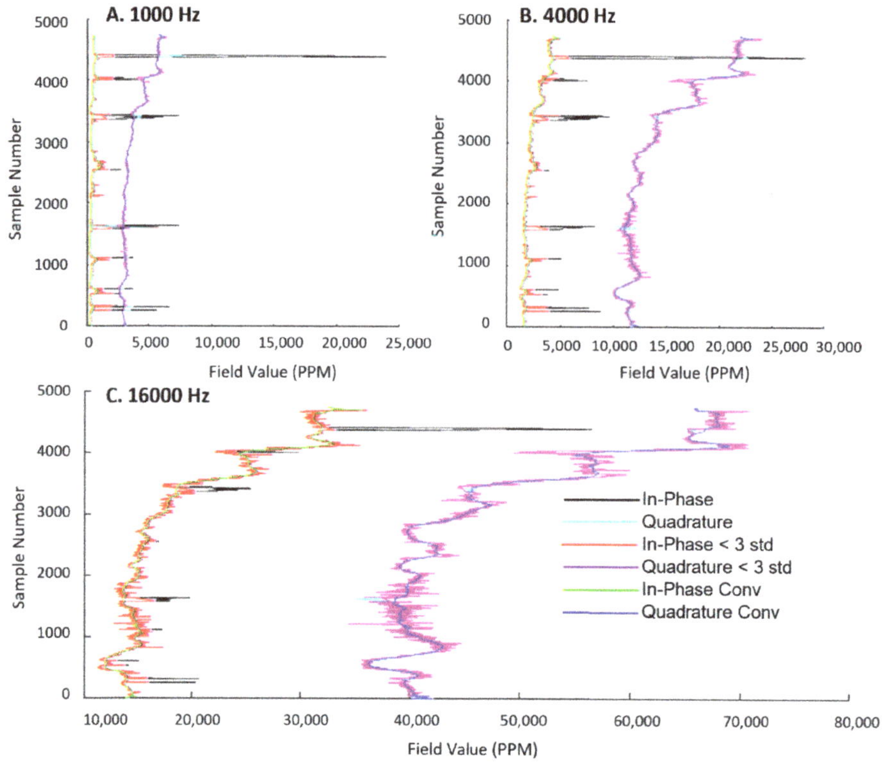

Figure 3. EM data reduction and smoothing in Seven Palm system on 6 April 2019 at (**A**). 1 KHz, (**B**). 4 KHz and (**C**) 16 KHz. The black and aqua colors indicate the in-phase and quadrature raw EM data. The red and pink colors indicate the In-phase and quadrature raw EM data after outliers were removed. The lime and blue colors indicate the in-phase and quadrature filtered EM data.

In the Seven Palm system, on 10 July 2019, high winds caused considerable chops in the water which introduced noise into the resistivity measurements associated with the larger spacings (n = 10 and 11). Those outlying datapoints were masked from the analyses. The average value of the measurements and reciprocals was used in the inversion. Water depth in the model was constrained to a rod measurement at the array center, but the water layer and lake bottom layer resistivity were left unconstrained in the inversion. Finally, the inverted resistivity was converted to salinity using the general equation defined in the Practical Salinity Scale of 1978 (S_p) [32–34]

$$S_p = \sum_{i=0}^{5} a_i (R_t)^{i/2} + \frac{(t_{68}\ /°\mathrm{C} - 15)}{[1 + 0.0162\ (t_{68}\ /°\mathrm{C} - 15)\ ]} + \sum_{i=0}^{5} b_i\ (R_t)^{i/2} \tag{2}$$

where $(t_{68}\ /°\mathrm{C})$ is the measured temperature, r is the measured conductivity in (μS/cm)/ 42,914, and R_t is the conductivity ratio r/r_t. The factor r_t is given by

$$r_t = \sum_{i=0}^{4} c_i (t_{68}\ /°\mathrm{C})^i \tag{3}$$

and the coefficients $a_i,\ b_i, and\ c_i$ are provided in Table 3 below. The salinity equation is approximately valid for salinity values that range between 2 to 42 PSU, where the temperature is between −2 to 35 °C [32–34].

Table 2. Descriptive statistical analysis of data reduction and smoothing: EM data of less than 3 standard deviations were removed and filtered by smoothing 100 consequent samples. The statistical analysis shows the result of 100 consequent samples of EM data from each survey date.

Site	Date	Frequency (KHz)	EM Observed Data	EM Data < 3 std	EM Data < 3 std and Smoothing	Data Removed	N Total	Mean	Standard Deviation	SE of Mean	Variance	Skewness	Kurtosis	Minimum	Median	Maximum
West Lake	January 2016	1	8923	8899	8872	51	100	1489	6.1	0.6	37	−0.50	−0.26	1473	1490	1498
West Lake	January 2016	4	8923	8899	8872	51	100	1595	5.6	0.6	32	−0.95	1.03	1578	1596	1604
West Lake	January 2016	16	8923	8899	8872	51	100	1785	7.5	0.8	56	0.32	0.06	1769	1784	1803
West Lake	February 2017	1	6248	6176	6077	171	100	1420	75.7	7.6	5733	0.46	−1.38	1336	1391	1565
West Lake	February 2017	4	6248	6176	6077	171	100	1529	87.0	8.7	7576	0.41	−1.48	1432	1493	1690
West Lake	February 2017	16	6248	6176	6077	171	100	1726	102.3	10.2	10,459	0.44	−1.48	1614	1687	1907
Seven Palm	February 2016	1	4012	4010	3998	14	100	1547	24.7	2.5	608	0.88	0.98	1499	1544	1624
Seven Palm	February 2016	4	4012	4010	3998	14	100	1630	28.6	2.9	819	0.83	0.82	1571	1625	1714
Seven Palm	February 2016	16	4012	4010	3998	14	100	1856	39.8	4.0	1581	0.08	−0.08	1774	1856	1955
Seven Palm	March 2016	1	6380	6357	6346	34	100	2027	50.3	5.0	2530	0.56	−0.98	1950	2012	2126
Seven Palm	March 2016	4	6380	6357	6346	34	100	2226	57.7	5.8	3332	0.57	−0.91	2139	2213	2342
Seven Palm	March 2016	16	6380	6357	6346	34	100	2815	90.4	9.0	8171	0.56	−0.88	2671	2796	2997
Seven Palm	April 2016	1	4710	4659	4555	155	100	1173	24.5	2.5	602	0.49	−0.28	1119	1171	1231
Seven Palm	April 2016	4	4710	4659	4555	155	100	1241	23.8	2.4	566	0.89	0.34	1205	1237	1309
Seven Palm	April 2016	16	4710	4659	4555	155	100	1387	23.8	2.4	568	0.53	−0.19	1342	1383	1453
Seven Palm	January 2017	1	5709	5709	5709	0	100	1124	20.1	2.0	404	1.47	2.16	1093	1117	1186
Seven Palm	January 2017	4	5709	5709	5709	0	100	1194	23.4	2.3	545	1.67	2.62	1165	1186	1268
Seven Palm	January 2017	16	5709	5709	5709	0	100	1282	22.2	2.2	491	1.14	1.01	1258	1279	1346

Table 3. The coefficients used to calculate salinity.

i	a_i	b_i	c_i
0	0.008	0.0005	6.766097×10^{-1}
1	−0.1692	−0.0056	2.00564×10^{-2}
2	25.3851	−0.0066	1.104259×10^{-4}
3	14.0941	−0.0375	-6.9698×10^{-7}
4	−7.0261	0.0636	1.0031×10^{-9}
5	2.7081	−0.0144	

The formation factor for the study area was estimated using the bulk resistivity measured by the floating VES and coincident resistivity (inversely proportional to conductivity) data measured in the groundwater wells. The average daily conductivity (μS/cm) measurement in the groundwater wells was converted to resistivity, where the resistivity (Ω·m) = 10,000/Conductivity water (μS/cm). Then, the formation factor was calculated from the average value of the ratio of bulk resistivity modeled from the geophysical data and the pore water resistivity measured in the shallow groundwater wells. This estimated formation factor was used to produce a regional map of groundwater salinity.

4. Results

4.1. Electromagnetism

4.1.1. Spatio-Temporal EM Apparent Conductivity Changes during the 2016 Dry Season in the Seven Palm System

In the Seven Palm system at monthly intervals during the 2016 dry season, the apparent conductivity measurements increased from February to April (Figure 4). The 1 KHz apparent conductivity ranged from 800 mS/m at the northern end of Seven Palm Lake to 1400 mS/m at southern end in February 2016, whereas in April 2016, it ranged from 930 to 2150 mS/m. Similarly, the apparent conductivity along 4 KHz ranged from 900 to 1500 mS/m in February 2016 to a range of 1000 to 2800 mS/m in April 2016 (Figure 4). This showed that the apparent conductivities increased from north to south at all frequencies. In addition, for example, in northwest Seven Palm Lake, the 1 KHz apparent conductivities increase from around 800 mS/m in February to around 1070 mS/m in March to over 1100 mS/m in April. In Middle Lake, the apparent conductivities increased from around 1500 mS/m in February to 1700 mS/m in March to over 1800 mS/m in April. In Monroe Lake, it increased from around 2300 mS/m in March to over 2500 in April. This trend is consistent for 4 KHz data as well.

The apparent conductivity of the 16 KHz data varies from 1100 to 2000 mS/m in February 2016 and from 1200 to 2400 mS/m in March 2016 and from 1400 to 3200 mS/m in April 2016 from north to south along the profile line. This is consistent with the increase in conductivity and salinity seen in the surface probe measurements.

In general, during the 2016 dry season, the median apparent conductivity measurements increased from February to April at all frequencies as shown in the boxplot (Figure 5). In the boxplot, the interquartile range (IQR) of the March 2016 data is highly dispersed compared to the other EM data set, because the survey line covered a larger portion of Monroe Lake (Figure 5), which is highly influenced by Florida Bay (Figure 1).

4.1.2. Spatio-Temporal EM Apparent Conductivity Changes between 2016 and 2017 in the Coastal Lakes of ENP

The EM data show a considerable decrease in apparent conductivity between February 2016 and January 2017 in both the Seven Palm and West Lake systems. In the Seven Palm system, the apparent conductivity showed a general decrease of 150 mS/m in the Seven Palm system and 300 mS/m in the Middle Lake (Figure 6A,B). In the Seven Palm system, the 1 KHz apparent conductivity ranged from 800 mS/m at the northern end of Seven Palm

Lake to 1400 mS/m at the southern end in February 2016 (Figure 6A), whereas in January 2017, it ranged from 740 to 1200 mS/m (Figure 6B). Similarly, the apparent conductivity along 4 KHz ranged from 870 to 4000 mS/m in 2016 (Figure 6C) to a range of 800 to 2400 mS/m in 2017 (Figure 6D).

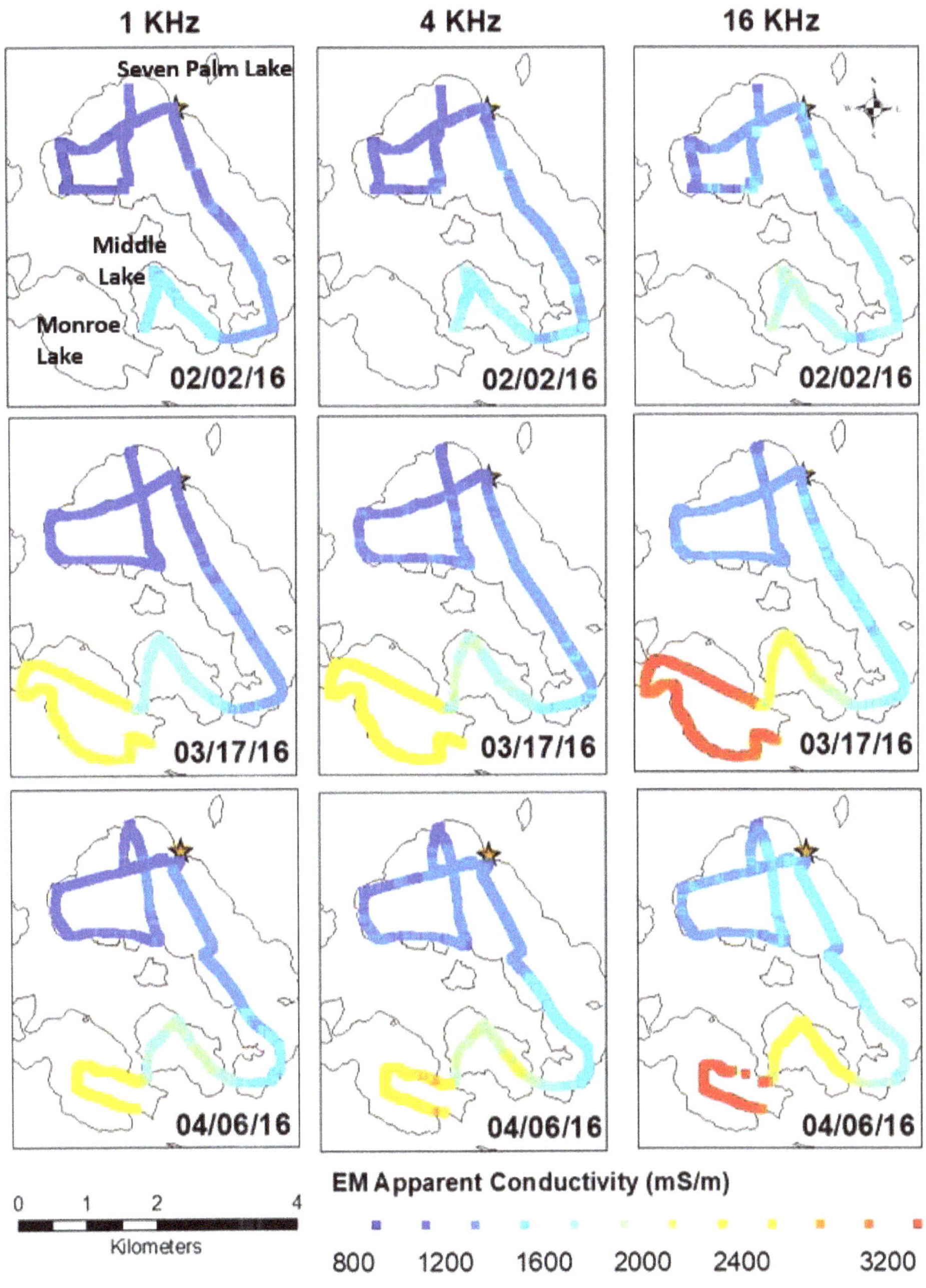

Figure 4. Map of apparent conductivities at 1, 4, and 16 KHz in 2016 in the Seven Palm system, Middle Lake and Monroe Lake. The star shows the location of the monitoring well.

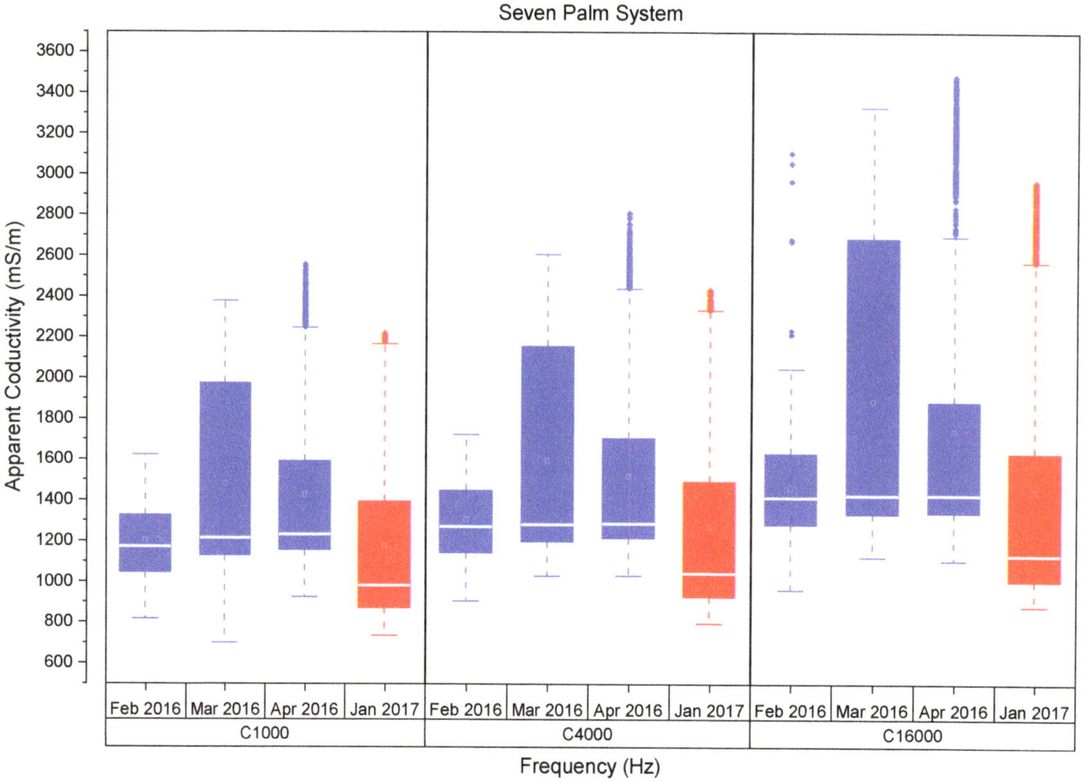

Figure 5. The box plot shows the mean (white dot), the median (white line), the interquartile range (IQR blue or red box) (i.e., the range between the 25th and 75th percentile), the upper and lower fence (range within 1.5 IQR), and outliers (asterisks). The EM data are not normally distributed, and the outliers (greater than 3 standard deviations of the consequent 100 samples) of the EM data were removed before the exploratory data analysis (boxplot). We considered the outliers marked by asterisks as field values.

The EM results show the same trend in the West Lake system. The apparent conductivity of the 1 KHz data decreases from a range of 1200 to 1700 mS/m in 2016 to 950 to 1850 mS/m in 2017 (Figure 6A,B). In addition, the result along 4 KHz dropped from 1600 mS/m in 2016 to 1500 mS/m in 2017 (Figure 6C,D). Similarly, the result along 16 KHz dropped from 1800 mS/m in 2016 to 1600 mS/ in 2017 (Figure 6E,F). In the West Lake system in January 2016, the highest apparent conductivities (2000 mS/m) were seen at the eastern end of the Long Lake and southern West Lake with the lowest conductivities in northern Cuthbert Lake.

In the West Lake system, the mean, median, and IQR of the apparent conductivity measurements decreased between February 2016 and January 2017 (Figure 7). Similarly, the median apparent conductivity measurements decreased between February 2016 and January 2017 in the Seven Palm system (Figure 5).

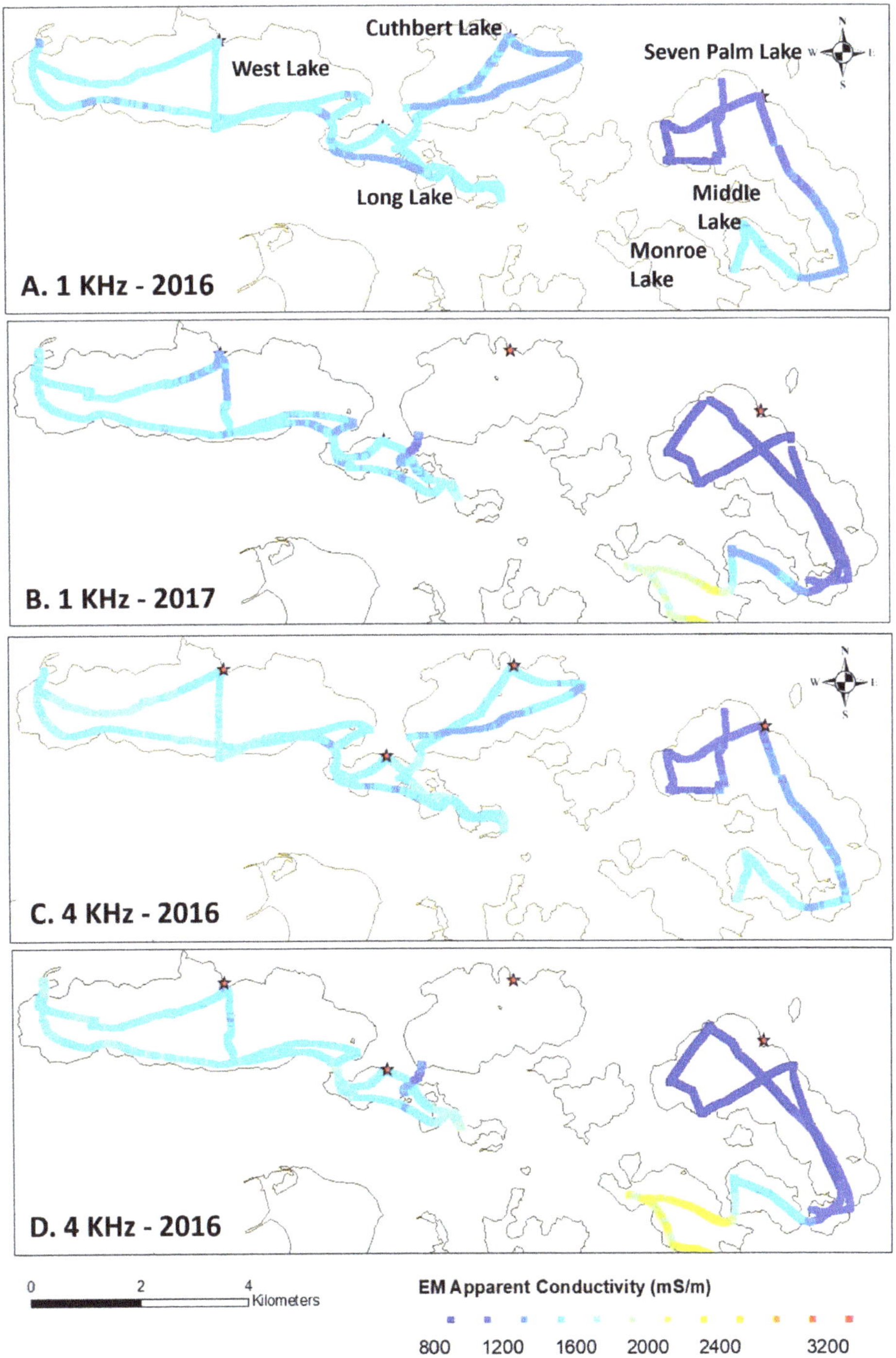

Figure 6. *Cont.*

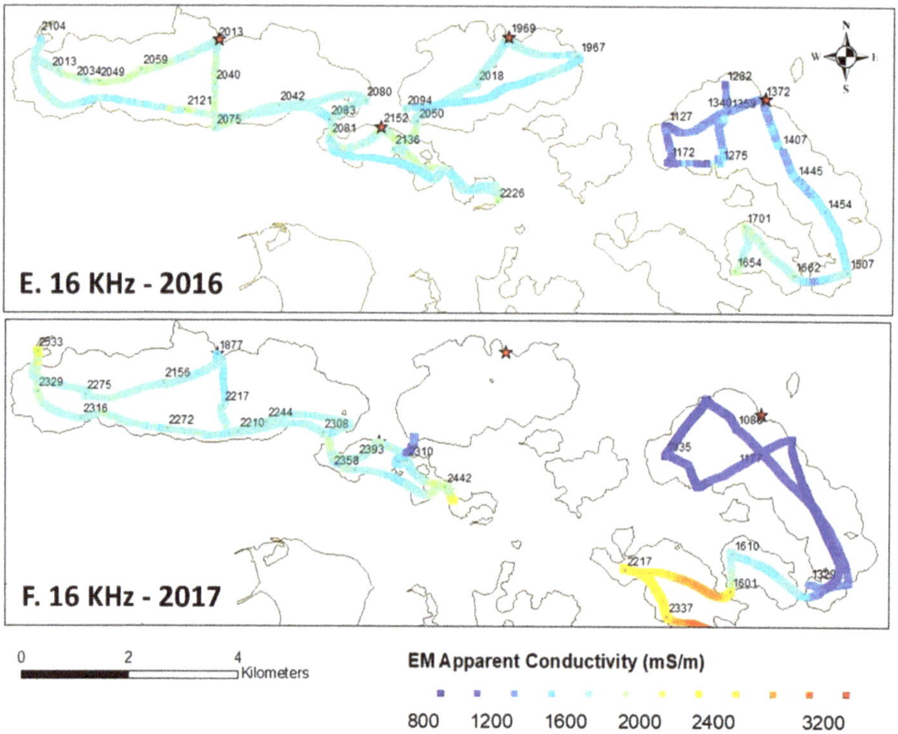

Figure 6. Map of apparent conductivities at West Lake and Seven Palm system: (**A**) 1 KHz in 2016, (**B**) 1 KHz in 2017, (**C**) 4 KHz in 2016, (**D**) 4 KHz in 2017, (**E**) 16 KHz in 2016, and (**F**) 16 KHz in 2017. At West Lake, the data were collected on 25 January 2016 and 10 February 2017. At Seven Palm, the data were collected on 6 February 2016 and 20 January 2017. The star shows the location of the monitoring well.

The apparent conductivity change between 2016 and 2017 in Seven Palm system is more significant than the West Lake system. The apparent conductivity change between 2017 to 2016, at all frequencies, decreased on average by 30% in Seven Palm system and 10% in West Lake system (Figure 8). At Seven Palm, the apparent conductivity change decreased by 20% at the northern end of Seven Palm Lake near the well and 30% at the southern end and clearly showed the trend from north to south. However, at the northern tip of Seven Palm there was an increase from 30 to 40%. Similarly, in the West Lake system, there was a gradual change from north to south and a significant change of 30% observed on the northern tip of Long Lake. In addition, the apparent conductivity change of the surface water is significant compared to the groundwater. For example, near the well at 1 KHz, 4 KHz, and 16 KHz, the apparent conductivity change decreased by 18%, 21%, and 26%, respectively (Figure 8A–C).

4.2. Electrical Resistivity

The resistivity data collected on 10 July 2019 in Seven Palm Lake and on 15 July 2019 in West Lake were inverted to a two-layer constrained water depth model. In this section, we focus on the spatial−temporal change in groundwater and surface water.

In the constrained water depth inversion, the best fit (red color) and Occam's inversion (olive color) showed a good agreement in estimating the water column resistivity measured using the YSI probe (Figure 9). In the figures, the equivalence analysis (dashed green line) produced a more tightly constrained result in the West Lake system (Figure 9a–j) compared to the Seven Palm system (Figure 9k–r). In the West Lake system, the surface water resistivity measured using the YSI probe decreased between 0.37 Ω·m and 0.28 Ω·m from west to east. In the Seven Palm system, the resistivity varies from 0.27 Ω·m at the north end to 0.17 Ω·m at the south end (Figures 9a–j and 10A). Similarly, the inverted surface water resistivity in the West Lake system decreased between 0.33 Ω·m at the west end and 0.26 Ω·m at the east end. In the Seven Palm system, the resistivity varies from 0.23 Ω·m at the north end to 0.16 Ω·m at the south end (Figures 9k–r and 10B). Generally, the data showed west to east and north to south decreases in surface water resistivity.

The lake bottom resistivity in the West Lake system varies between 1.69 Ω·m at the west end and 3.74 Ω·m at the east end. In the Seven Palm system, the resistivity is 32.85 Ω·m in the north and decreases gradually toward the south to 1.5 Ω·m (Figure 10C). The lake bottom resistivity generally increased from west to east while it decreased from north to south.

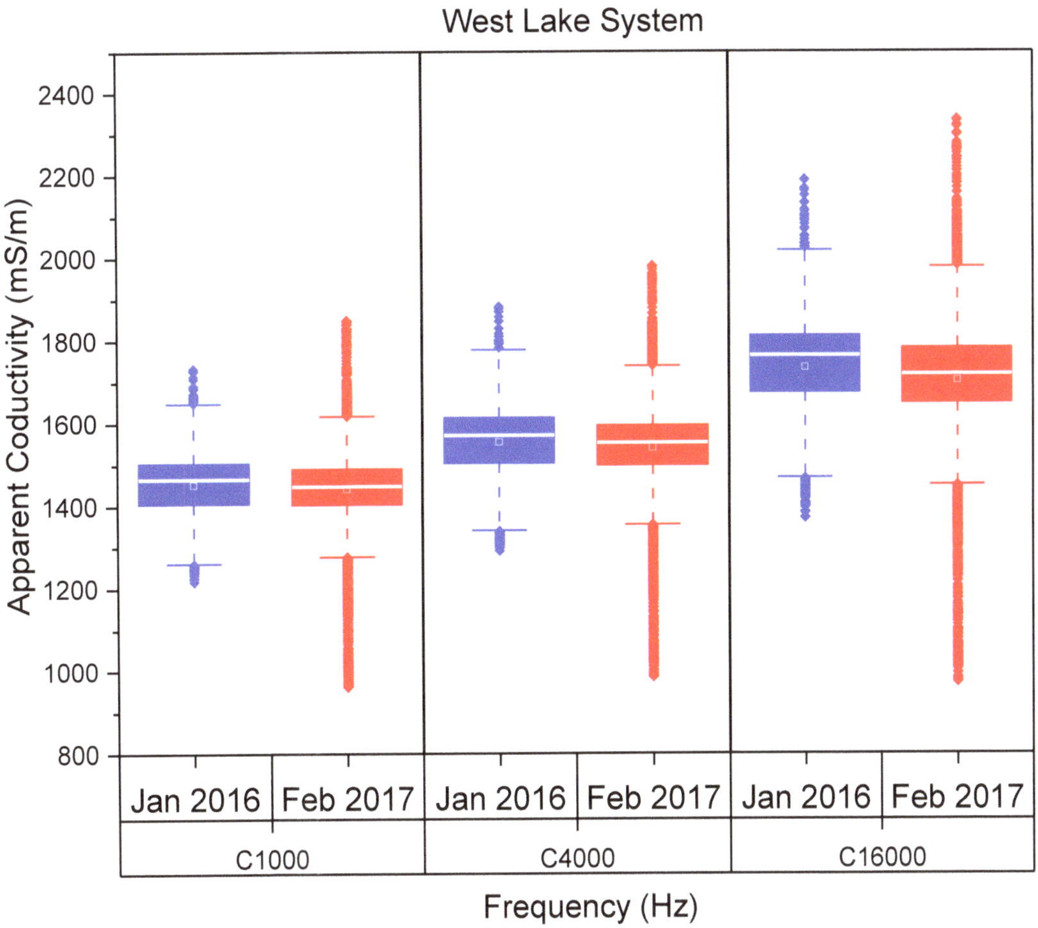

Figure 7. The box plot shows the mean (white dot), the median (white line), the interquartile range (IQR blue or red box) (i.e., the range between the 25th and 75th percentile), the upper and lower fence (range within 1.5 IQR), and outliers (asterisks).

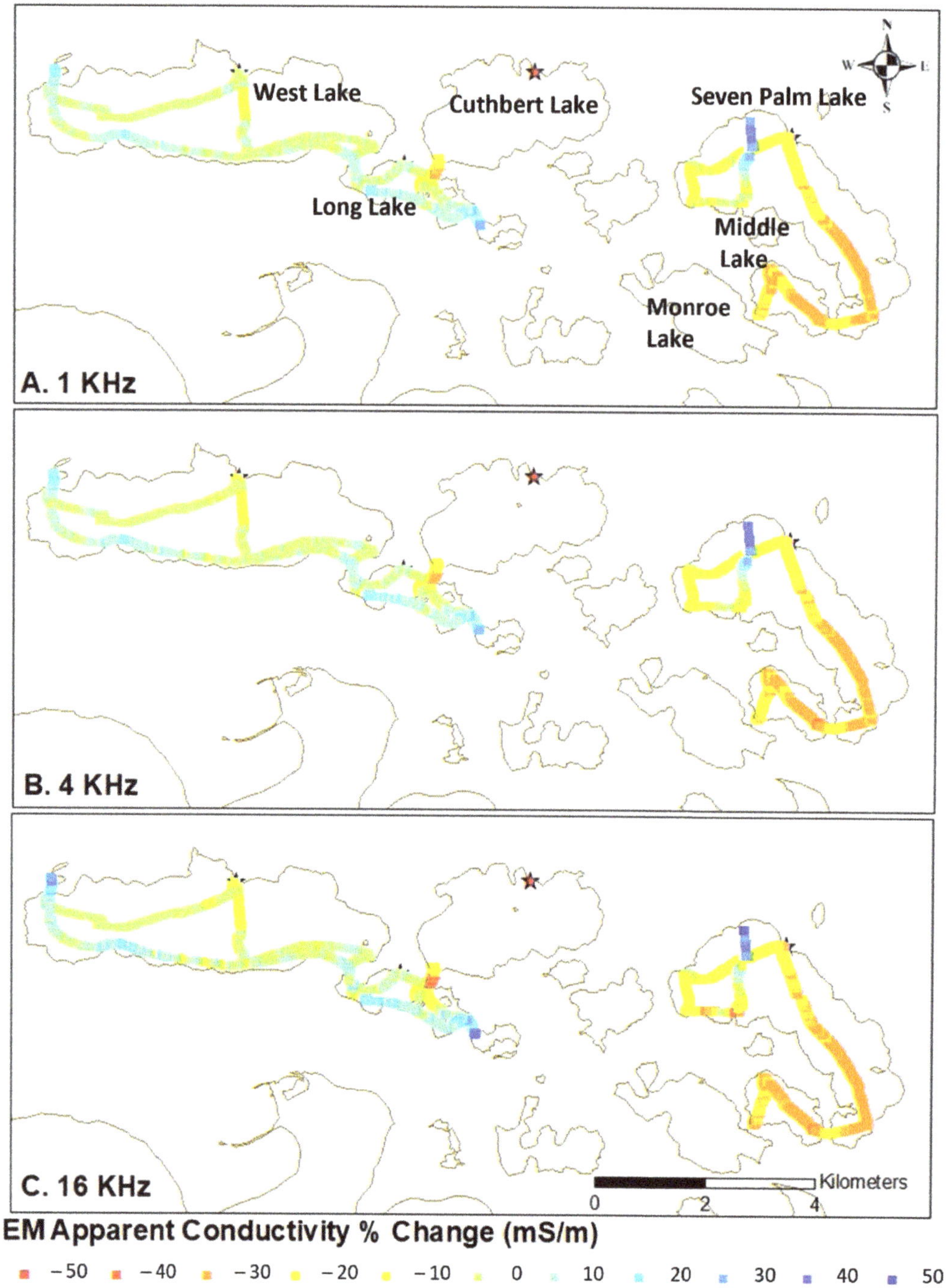

Figure 8. Apparent conductivity percent change between 2017 and 2016 in the West Lake and Seven Palm systems. The star shows the location of the monitoring well.

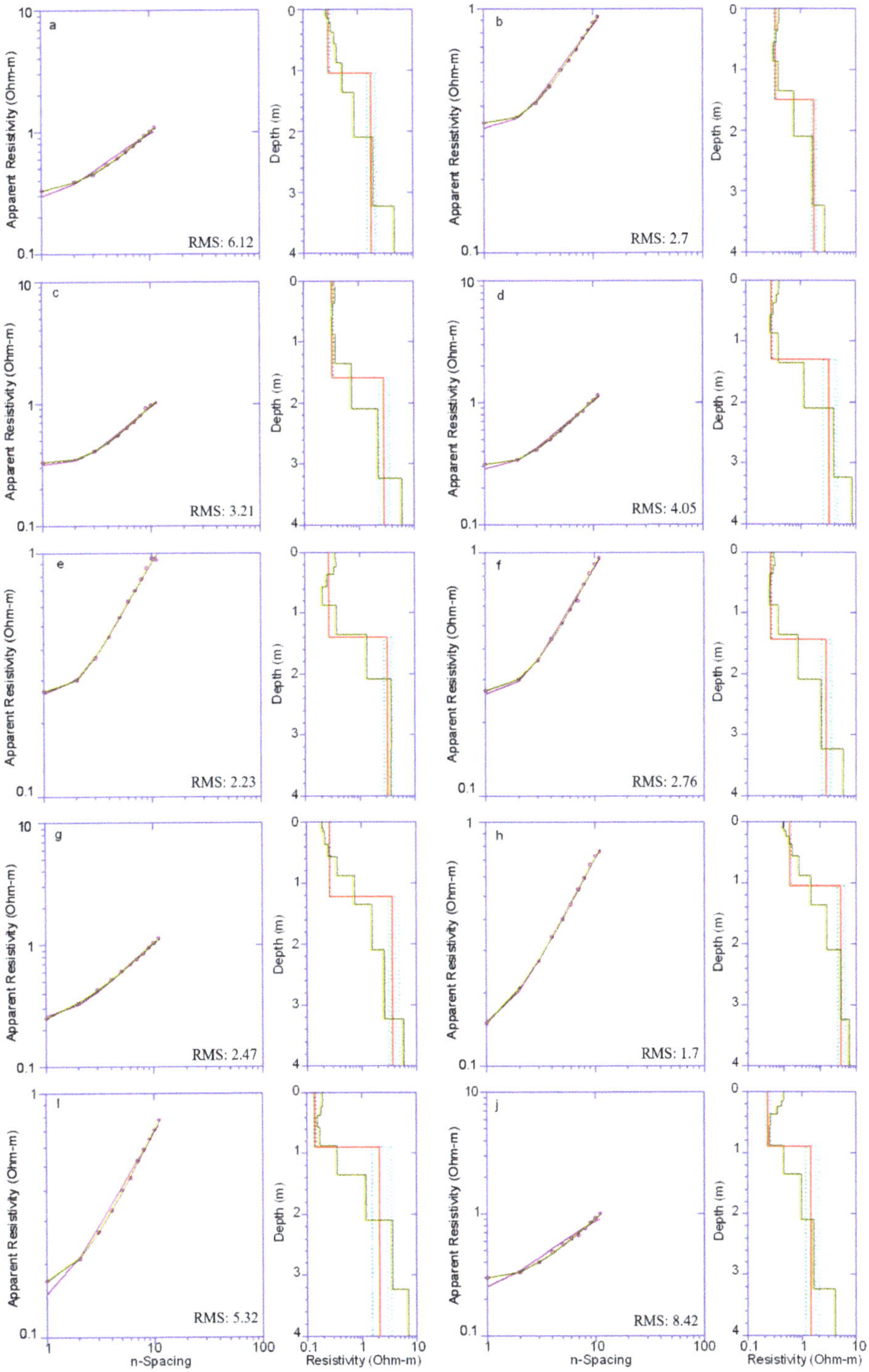

Figure 9. *Cont.*

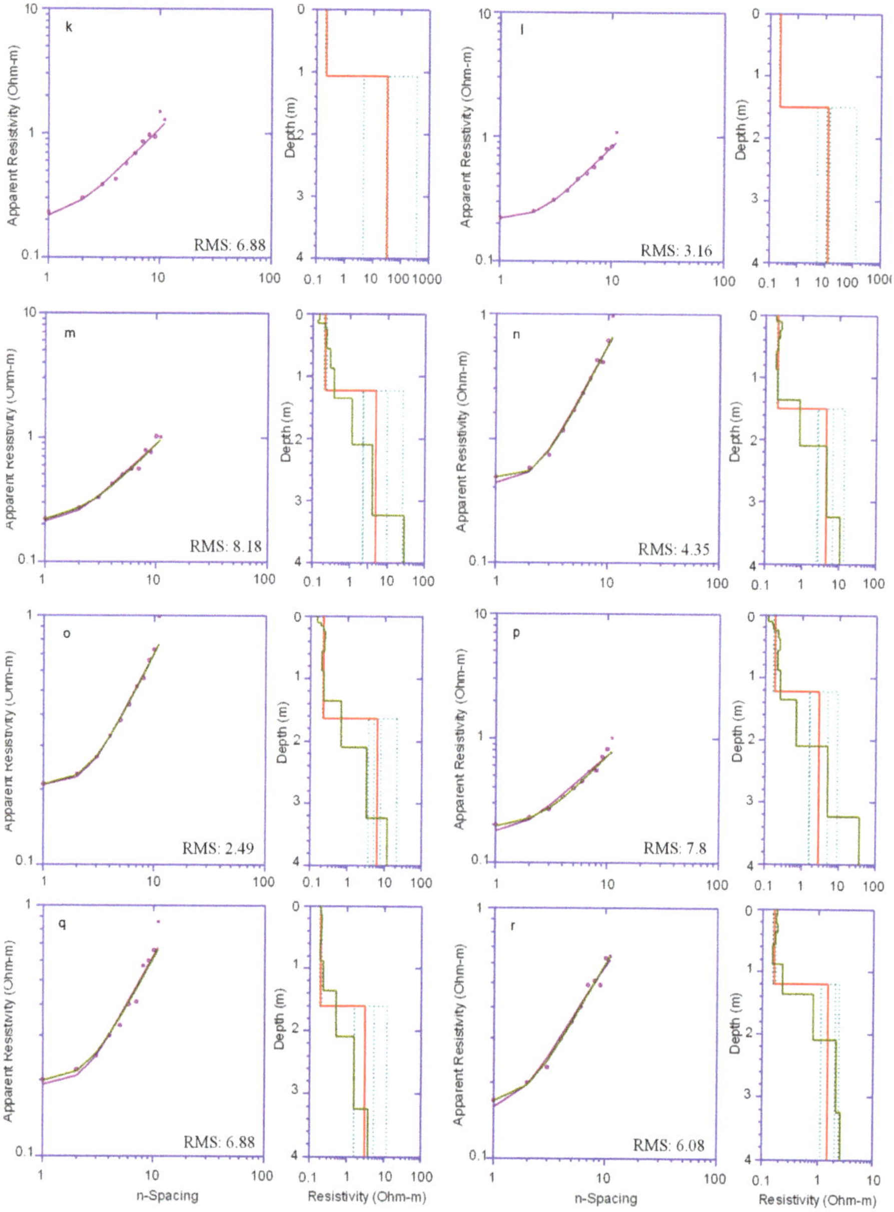

Figure 9. Constrained resistivity inverse model in West Lake system (**a**–**j**) and Seven Palm system (**k**–**r**). The left panel shows the observed and calculated data (lines), and the right panel shows the inverted model. In the left panel, the square dots indicate the observed data, the purple line indicates the best fit calculated data, and the olive color indicates Occam's inversion. In the right panel, the red line indicates the best fit model, the green dashed lines indicate the equivalency analysis, and the olive line represents Occam's inversion model. The outlier measurements represented by asterisks were masked during inversion.

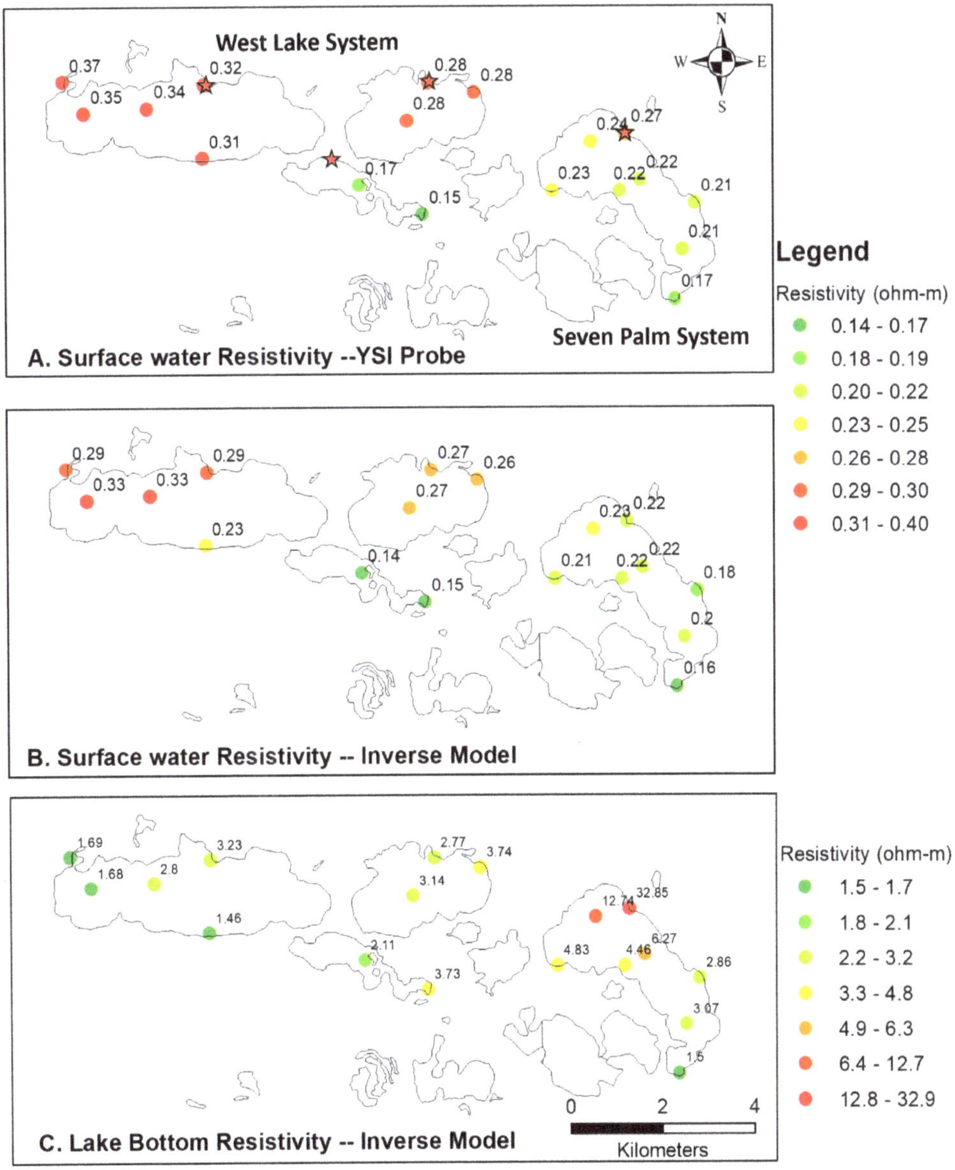

Figure 10. Resistivity in West Lake and Seven Palm. (**A**) Surface water using YSI probe. (**B**) Surface water from the inverse model. (**C**) Lake bottom resistivity from the inverse model. The star shows the location of the monitoring well.

In the inverse model, the minimum and maximum lake bottom resistivity varied between 1.4 Ω·m and 3.7 Ω·m at West Lake and between 1.4 Ω·m and 32.8 Ω·m at Seven Palm. Based on equivalency analysis, at West Lake, the model showed a highly constrained model with the lower and upper limit between 1.1 Ω·m and 4.9 Ω·m (Figure 11), whereas in Seven Palm, the model showed a poorly constrained model with the lower and upper limit between 1.4 Ω·m and 362 Ω·m (Figure 11). This model is poorly constrained mainly on the upper bound. The lower and upper limits in West Lake are highly constrained compared with the Seven Palm system (Figure 11). This uncertainty on the upper bound could be due to measurement errors induced by wind or wave action.

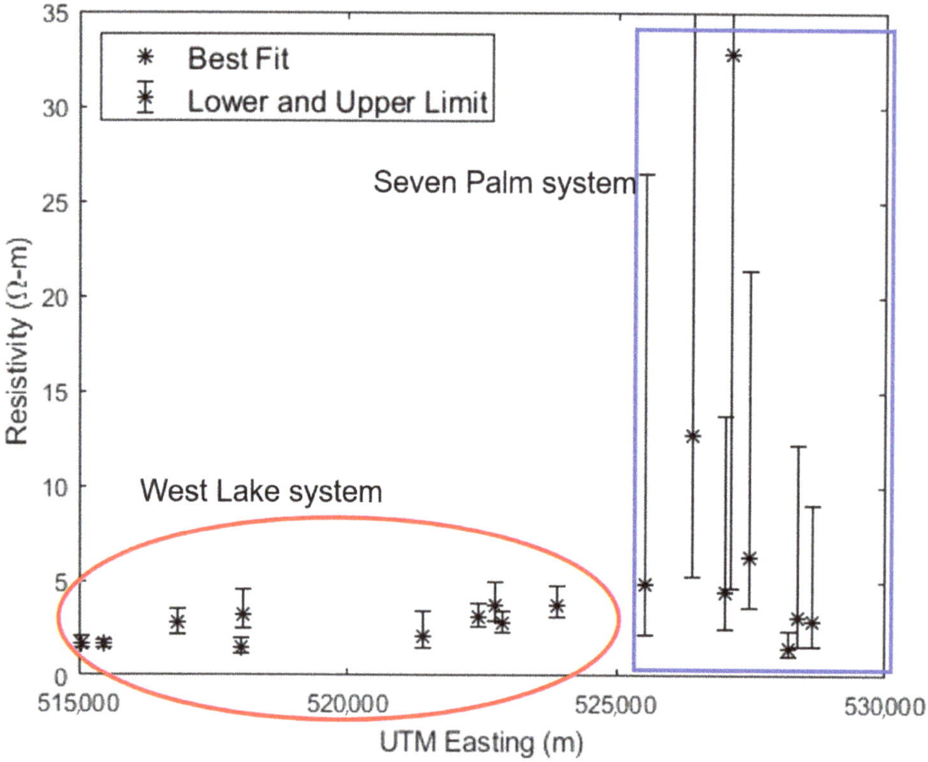

Figure 11. West–east profile showing the inverted lake-bottom resistivity from the VES surveys. Error bars indicate the lower and upper acceptable limits derived from equivalence analysis. West Lake system is located between 515,000 and 525,000 m and the Seven Palm system is located between 525,000 and 530,000 m.

4.2.1. Formation Factor

One method of estimating the formation factor of the lake bottom formation is by comparing groundwater resistivity (ρ_w) measured in the shallow groundwater wells with nearby values of lake bottom resistivity $\left(\rho_f\right)$ modeled from the geophysical data. In this study, the West Lake data collected on 15 July 2019, was used to calculate a formation factor for the coastal lakes of ENP. A mean local formation factor of 10.7 ± 1.8 is estimated for the Mangrove Lakes of ENP. This estimated formation factor is calculated from the average formation factor of West Lake, Long Lake, and Cuthbert Lake (Table 4). The formation factor of Seven Palm was not considered in the calculation as the resistivity of the model was poorly constrained due to the noise in the data (Figure 9).

4.2.2. Salinity

The surface water salinity measured using the YSI probe varies from 15 PSU in the west in West Lake to 21 PSU in the east in Seven Palm (Figure 12A). In the Seven Palm system, the surface water salinity measured using the YSI probe varies from 21 PSU in the north to 33 PSU in the south (Figure 12A).

Table 4. Formation factors in the coastal lakes of ENP.

Site	Date	Distance from the Well (m)	Well Data		Geophysical Model		F
			Surface Water	Groundwater ρ_w	Surface Water	Lake Bottom ρ_f	
West Lake	15 July 2019	68	0.32	0.26	0.29	3.23	12.6
Long Lake	15 July 2019	807	0.19	0.20	0.14	2.11	10.6
Cuthbert Lake	15 July 2019	48	0.28	0.31	0.27	2.77	9.0
Seven Palm	10 July 2019	41	0.23	1.12	0.22	32.85	29.41

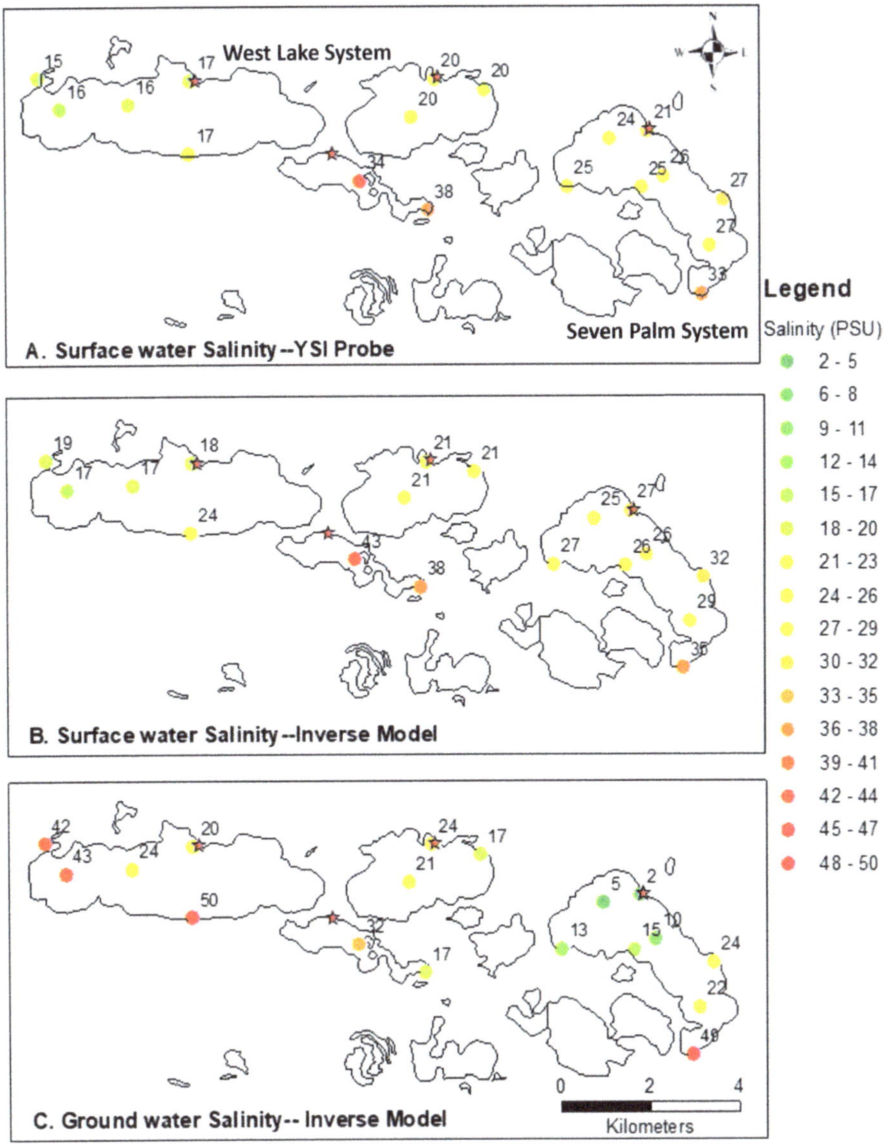

Figure 12. Salinity in West Lake and Seven Palm. (**A**) Surface water using YSI probe. (**B**) Surface water from the inverse model. (**C**) Groundwater from the inverse model.

The inverted surface water salinity has the same trend and increased from 19 PSU to 27 PSU from west to east and from 27 PSU to 35 PSU from north to south (Figure 12B). In contrast, the inverted groundwater salinity decreased from 42 PSU in the west in West Lake to 2 PSU in the east in Seven Palm. In the Seven Palm system, the groundwater salinity increased from 2 PSU in the north to 49 PSU in the south (Figure 12C).

5. Discussion

During the 2016 dry season, the apparent conductivity measurements at 1 KHz and 4 KHz increased from February to April (Figures 4 and 5) in the Seven Palm system at monthly intervals. This reflects a general increase in apparent conductivities (salinity) in the groundwater over the dry season. This change is in response to decreased precipitation (Figure 13). Similarly, the apparent conductivity of the 16 KHz data indicates a general increase in apparent conductivities (salinity) in the surface water over the dry season and from north to south along the profile line of the Seven Palm system. This change is due to reduced precipitation and increased potential evapotranspiration, which results in a sharp decrease in the water level as well (Figure 13). The surface water salinities of the coastal lakes of ENP showed brackish to hypersaline salinities with higher salinities in the summer following the dry season [22,23]. Salinity variation in time and space in the Florida Bay is governed by the influence of precipitation [35] evapotranspiration, runoff, mass exchange with the surrounding basins [5], proximity of upstream freshwater sources, and exchange with Florida Bay marine waters downstream [22,23].

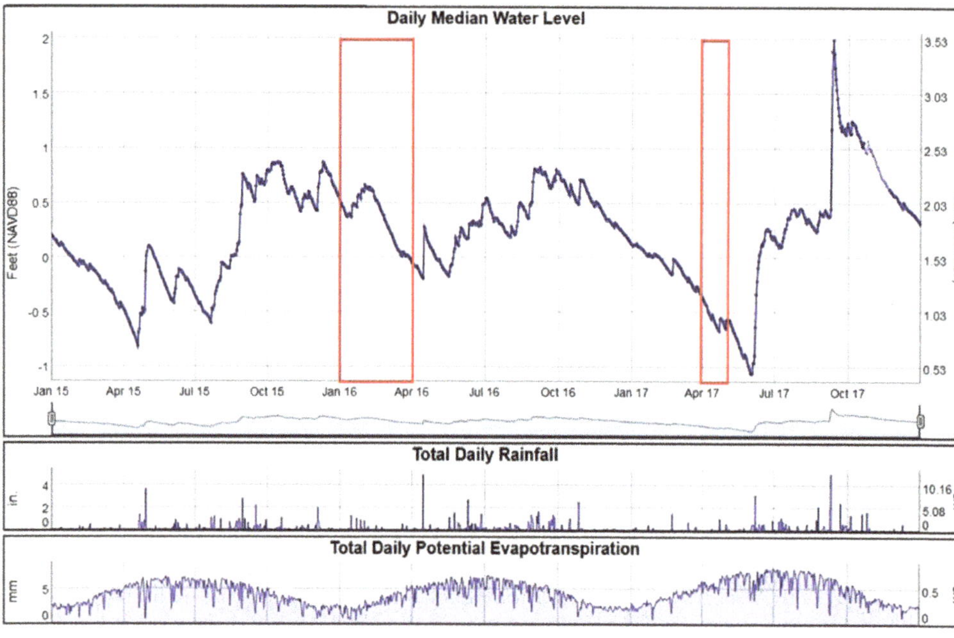

Figure 13. Water level, rainfall, and evapotranspiration data between January 2015 and December 2017 at the NCL station, Everglades National Park. The red rectangle indicates the decrease in rainfall during the EM survey period in 2016 and 2017. The data source is the Everglades Depth Estimation Network (EDEN), USGS website http://sofia.usgs.gov/eden/station.php (accessed on 15 July 2020).

Between February 2016 and January 2017, the EM data show a considerable decrease in apparent conductivity in both the Seven Palm and West Lake system. A general decrease of 150 mS/m in Seven Palm system and 300 mS/m in Middle Lake (Figure 6a,b) suggests a freshening of the groundwater and may reflect the effects of increased flow due to restoration efforts. Lower conductivities observed in Cuthbert Lake compared to West Lake and Long Lake may reflect increases in freshwater flows from upstream sources, possibly from increased flows due to restoration efforts.

The apparent conductivity change between 2016 and 2017 in the Seven Palm system is more significant than the West Lake system. The apparent conductivity change decreased on average by 30% in the Seven Palm system and 10% in the West Lake system (Figure 8) at all frequencies. In general, in the coastal lakes of ENP, the decrease in apparent conductivity change is pronounced from west to east at all frequencies. This suggests the restoration efforts may have made a more observable contribution to the Seven Palm system but not yet in the West Lake system. Similarly, other studies conducted by the U.S. Geological Survey to map saltwater intrusion using helicopter electromagnetic (HEM) surveys identified variations in the apparent conductivity that were associated with changes in groundwater quality [39]. The apparent resistivity map (inversely related to conductivity) from the HEM survey showed a transition from low to high resistivity from West Lake to Seven Palm, i.e., high to low conductivity from west to east, which is consistent with the results of this study. The authors concluded the change is due to the influence of freshwater flowing in the Taylor Slough which pushes the transition in a southwesterly direction in the area of maximum flow [40]. A mean local formation factor of 10.7 ± 1.8 is estimated for the Mangrove Lakes of ENP. This estimated formation factor is similar to previously published estimates of formation factor 9.65 for the ENP [19]. We used the estimated formation factor 10.7 in this study to calculate the salinity of the groundwater. This approach has the advantage of forcing the geophysics derived salinities to be consistent with the in situ well measurements.

The surface water salinity increased from west to east, and the groundwater salinity decreased from west to east. In another study conducted by [41], in the Seven Palm system in July 2015, the surface water salinity ranged from 25 PSU in the northern tip of Seven Palm and gradually increased to 45 PSU in Monroe Lake. These hyper-salinity conditions are associated with drought and produced seagrass die-off [8,41]. In July 2019, the surface water salinity ranged from 5 PSU in the northern tip of Seven Palm and gradually increased to 49 PSU in Monroe Lake. Generally, in the West Lake system, the surface water salinity is lower than the groundwater salinity, but in the Seven Palm system, the surface water salinity is higher than the groundwater salinity (Figure 12A). In addition, in both lakes, the salinity of the surface water and groundwater increased from north to south. This result suggests a freshening of the groundwater and may reflect the effects of increased flow due to restoration efforts.

In the Mangrove Lakes of ENP, surface water salinity is driven by seasonally variable winds and evaporation, whereas groundwater salinity is less temporally variable and reflects regional groundwater flow. In the EM data, the apparent conductivity change of the surface water is significant compared to the groundwater (Figure 7). In the VES data, viewed in a west to east profile, less saline surface water in West Lake overlies more saline groundwater (Figure 11). In contrast, in Seven Palm Lake, the vertical salinity profile is inverted with denser, more saline surface water overlying less dense, fresher groundwater. This inverted salinity profile is gravitationally unstable and likely has a profound effect on recharge and discharge mechanisms to and from the aquifer. These results suggest the floating electrical resistivity array can map the resistivity of the subsurface formation effectively.

Some limitations of the estimated formation factor and EM measurements include the following. In the Mangrove Lakes of ENP, the shallow wells only extended to the base of the peat layer and did not penetrate the underlying limestone. This method assumes that the groundwater sampled from the wells is derived from the underlying limestone. Furthermore, repeated measurements would be useful to determine the accuracy of the estimated formation factor value with minimum standard deviation.

In the EM data, the apparent conductivities at certain locations where the EM survey line crosses have different measurements. This difference could be due to instrument drift, the operator changing position during the survey which affects the position of the coils, and calibration problems after Bluetooth connection failure which causes a new calibration in different environmental settings. Furthermore, the EM result is not inverted as the inverted model is not consistently reliable. As a result, the EM data are presented in apparent conductivity. The authors of [42] suggested that discrepancies in inversion for soils having a high surface electrical conductivity relative to deeper depths using EM 38 can be compensated through the vertical or horizontal coil configuration response curve. The EM method provides a faster alternative to DC soundings [16], but it is sensitive to electromagnetic noise produced by metals, pipelines, etc. An integrated approach of electrical resistivity and EM methods can produce the best result [43]. In a saline environment, once areas of high salinity are delineated, alternative methods of salinity measurement can be utilized to provide a detailed description of the study area [42]. Therefore, an integrated approach of using EM, resistivity, and other geological information is necessary to characterize the subsurface formation accurately.

6. Conclusions

Geophysical methods can map the difference in conductivities (salinity) of a subsurface formation effectively. The results of the EM survey show the apparent conductivity measurements systematically increased from February to April 2016 in the Seven Palm system. This demonstrates how the salinity of the surface water and groundwater changes during the dry season. The general increase in the salinity reflects a decreased precipitation, increased evapotranspiration, and the influence of saline water from Florida Bay on the surface water and groundwater. The apparent conductivity showed a considerable decrease between February 2016 and January 2017 in both the West Lake and Seven Palm systems. These changes suggest a freshening of the groundwater and may reflect the effects of increased flow due to restoration efforts. Furthermore, the apparent conductivity change between 2016 and 2017 is more pronounced in the Seven Palm system than in the West Lake system. This suggests the restoration efforts may have made a positive contribution to the Seven Palm system but not yet to the West Lake system.

The floating resistivity survey efficiently characterized the spatial variations in surface water groundwater salinity. In general, both the surface water and the groundwater increased in salinity from north to south. In contrast, the surface water salinity increased from west to east while the groundwater salinity decreased from west to east.

We estimated a mean local formation factor of 10.7 ± 1.8 for the coastal lakes of ENP. This estimated formation factor is slightly higher than the average regional formation factor of 9.65 estimated in another study [18]. The fact that the formation factor is obtained from few measurements may limit the finding of the salinity value in the coastal lakes of ENP. Hence, a further research study is needed to determine the accuracy of the estimated formation factor with minimum standard deviation. Another limitation of this study could be that the groundwater wells are situated approximately 1 m deep and may not represent the deeper groundwater.

Author Contributions: Conceptualization, M.E.K. and D.W.; methodology, M.E.K. and D.W.; software, M.E.K. and D.W.; validation, M.E.K., D.W., R.M.P., T.A.F. and C.J.M.; formal analysis, M.E.K.; investigation, M.E.K. and D.W.; resources, D.W.; data curation, M.E.K., D.W., R.M.P. and T.A.F.; writing—original draft preparation, M.E.K.; writing—review and editing, M.E.K., D.W., R.M.P., T.A.F. and C.J.M.; visualization, M.E.K.; supervision, D.W. and R.M.P.; project administration, D.W.; funding acquisition, R.M.P. All authors have read and agreed to the published version of the manuscript.

Funding: This research was funded by the South Florida Water Management District, Contract numbers PO9500005699 and PO4500093831. This is contribution number 1401 from the Institute of Environment at Florida International University.

Data Availability Statement: Data will be available to readers upon reasonable request.

Acknowledgments: We are grateful to Assefa Melesse, Pete Markowitz, and the anonymous reviewers for reviewing the paper and providing valuable suggestions that improved the manuscript. We are deeply grateful to Josh Allen, who contributed to the data acquisition.

Conflicts of Interest: The authors declare no conflict of interest. The funders had no role in the design of the study; in the collection, analyses, or interpretation of data; in the writing of the manuscript; or in the decision to publish the results.

References

1. Price, R.; Schwartz, K. Water, Sustainability, and Survival. In *The Coastal Everglades: The Dynamics of Social-Ecological Transformation in the South Florida Landscape*; Childers, D.L., Gaiser, E.E., Odgen, L.A., Eds.; Oxford University Press: Oxford, UK, 2019; pp. 34–70.
2. Price, R.M.; Top, Z.; Happell, J.D.; Swart, P.K. Use of tritium and helium to define groundwater flow conditions in Everglades National Park. *Water Resour. Res.* **2003**, *39*, 1–12. [CrossRef]
3. US Army Corps of Engineers (USACE) and South Florida Water Management District (SFWMD); Central and Southern Florida Comprehensive Review Study. Final Integrated Feasibility Report and Programmatic Environmental Impact Statement. 1999. Available online: https://www.sfwmd.gov/document/central-and-southern-florida-project-comprehensive-review-study-final-integrated (accessed on 21 August 2021).
4. Dessu, S.B.; Price, R.M.; Troxler, T.G.; Kominoski, J.S. Effects of sea-level rise and freshwater management on long-term water levels and water quality in the Florida Coastal Everglades. *J. Environ. Manag.* **2018**, *211*, 164–176. [CrossRef] [PubMed]
5. Kelble, C.R.; Johns, E.M.; Nuttle, W.K.; Lee, T.N.; Smith, R.H.; Ortner, P.B. Salinity patterns of Florida Bay. *Estuar. Coast. Shelf Sci.* **2007**, *71*, 318–334. [CrossRef]
6. Barlow, P.M. *Groundwater in Freshwater-Saltwater Environments of the Atlantic Coast*; United State Geological Survey (USGS): Menlo Park, CA, USA, 2003; Volume 1262. Available online: https://pubs.usgs.gov/circ/2003/circ1262/ (accessed on 23 September 2020).
7. McIvor, C.C.; Ley, J.A.; Bjork, R.D. Changes in freshwater inflow from the Everglades to Florida Bay including effects on biota and biotic processes: A review. In *Everglades: The Ecosystem and Its Restoration*; St. Lucie Press: Boca Raton, FL, USA, 1994; pp. 117–146.
8. NPS gov. Florida Bay Seagrass Die-Off National Park Service, U.S. Department of the Interior, South Florida Natural Resources Center Everglades National Park. 2015. Available online: https://www.nps.gov/ever/learn/nature/upload/seagrassDieoff_final_web_hi_res.pdf (accessed on 14 April 2020).
9. Jansen, J.R. Geophysical Methods to Map Brackish and Saline Water in Aquifers. Proceedings of the 2011 Georgia Water Resources Conference. 2011. Available online: http://smartech.gatech.edu/handle/1853/46031 (accessed on 14 April 2020).
10. Greenwood, W.J.; Kruse, S.; Swarzenski, P. Extending electromagnetic methods to map coastal pore water salinities. *Ground Water* **2006**, *44*, 292–299. [CrossRef]
11. Rajab, J.S.A.; El-Naqa, A.R. Mapping groundwater salinization using transient electromagnetic and direct current resistivity methods in Azraq Basin, Jordan. *Geophysics* **2014**, *78*, B89–B101. [CrossRef]
12. Mansoor, N.; Slater, L.; Artigas, F.; Auken, E. High-resolution geophysical characterization of shallow-water wetlands. *Geophysics* **2006**, *71*, B101–B109. [CrossRef]
13. Nenna, V.; Herckenrath, D.; Knight, R.; Odlum, N.; Mcphee, D. Application and evaluation of electromagnetic methods for imaging saltwater intrusion in coastal aquifers: Seaside Groundwater Basin, California. *Geophysics* **2013**, *78*, B77–B88. [CrossRef]
14. Machado, F.C.; Montanari, R.; Shiratsuchi, L.S.; Lovera, L.H.; Lima, E.D.S. Spatial Dependence of Electrical Conductivity and Chemical Properties of the Soil By Electromagnetic Induction. *Rev. Bras. Ciência Do Solo* **2015**, *39*, 1112–1120. [CrossRef]
15. US EPA. Resistivity Methods | Environmental Geophysics. 2021. Available online: https://archive.epa.gov/esd/archive-geophysics/web/html/resistivity_methods.html (accessed on 31 October 2021).
16. Corriols, M.; Ryom Nielsen, M.; Dahlin, T.; Christensen, N.B. Aquifer investigations in the León-Chinandega plains, Nicaragua, using electromagnetic and electrical methods. *Near Surf. Geophys.* **2009**, *7*, 413–426. [CrossRef]
17. Archie, G.E. The Electrical resistivity log as an aid in determining some reservoir characteristics. *Trans. AIME* **1942**, *146*, 54–62. [CrossRef]
18. Fitterman, D.V.; Prinos, S.T. *Results of Time-Domain Electromagnetic Soundings in Miami-Dade and Southern Broward Counties, Florida: U.S. Geological Survey*; Open-File Report; The United States Geological Survey (USGS): Reston, VA, USA, 2011; 289p. Available online: https://pubs.usgs.gov/of/2011/1299/ (accessed on 11 November 2021).
19. Fitterman, D.V.; Deszcz-Pan, M.; Stoddard, C.E. *Results of Time-Domain Electromagnetic Soundings in Everglades National Park, Florida (on CD–ROM): U.S. Geological Survey*; Open-File Report; U.S. Geological Survey | U.S. Department of the Interior: Reston, VA, USA, 1999; pp. 99–426. Available online: https://pubs.er.usgs.gov/publication/ofr99426 (accessed on 25 November 2021).
20. Tucker, N.M. *Analyzing Tidal Fluctuations in the Big Pine Key Freshwater Lens with Time-Lapse Resistivity*; Florida International University: Miami, FL, USA, 2013. [CrossRef]
21. Fitterman, D.V. Mapping Saltwater Intrusion in the Biscayne Aquifer, Miami-Dade County, Florida using Transient Electromagnetic Sounding. *J. Environ. Eng. Geophys.* **2014**, *19*, 33–43. [CrossRef]
22. Frankovich, T.A.; Morrison, D.; Fourqurean, J.W. Benthic macrophyte distribution and abundance in estuarine mangrove lakes: Relationships to water quality. *Estuaries Coasts* **2011**, *34*, 20–31. [CrossRef]

23. Frankovich, T.A.; Barr, J.G.; Morrison, D.; Fourqurean, J.W. Differing temporal patterns of Chara hornemannii cover correlate to alternate regimes of phytoplankton and submerged aquatic-vegetation dominance. *Mar. Freshw. Res.* **2012**, *63*, 1005–1014. [CrossRef]
24. GSSI. Profiler EMP-400. Available online: https://www.geophysical.com/products/profiler-emp-400 (accessed on 24 July 2021).
25. AGIUSA. SuperStingTM Wi-Fi. 2019. Available online: https://www.agiusa.com/supersting-wifi (accessed on 17 July 2020).
26. Binley, A.; Kemna, A. DC Resistivity and Induced Polarization Methods. In *Hydrogeophysics*; Springer: Berlin/Heidelberg, Germany, 2005; pp. 129–156. [CrossRef]
27. Binley, A. Tools and techniques: Electrical methods. In *Treatise on Geophysics*, 2nd ed.; Schubert, G., Ed.; Elsevier: Oxford, UK, 2015; pp. 233–259. [CrossRef]
28. Constable, S.C.; Parker, R.L.; Constable, C.G. Occam's inversion: A practical algorithm for generating smooth models from electromagnetic sounding data. *Geophysics* **1987**, *52*, 289–300. [CrossRef]
29. Auken, E.; Christiansen, A.V. Layered and laterally constrained 2D inversion of resistivity data. *Geophysics* **2004**, *69*, 752–761. [CrossRef]
30. Stoyer, C. 1D Resistivity, IP, EM Software (Version 3). 2019. Available online: http://www.interpex.com/ix1dv3/ix1dv3.htm (accessed on 17 July 2020).
31. Inman, J.R. Resistivity Inversion with Ridge Regression. *Geophysics* **1975**, *40*, 798. [CrossRef]
32. Unesco. *The Practical Salinity Scale 1978 and the International Equation of State of Seawater 1980. Unesco Technical Papers in Marine Science*; Unesco: Paris, France, 1981; 25p.
33. Unesco. *Algorithms for Computation of Fundamental Properties of Seawater. Unesco Technical Papers in Marine Science*; Unesco: Paris, France, 1983; 53p.
34. About Friends of Five Creeks. 2022. Available online: http://www.fivecreeks.org/monitor/sal.html (accessed on 2 January 2022).
35. Allen, J.M.; Price, R.M. Hydrologic Conditions in West and Seven Palm Lake Drainage Systems in the Florida Everglades. In Proceedings of the 2015 LTER All Scientists Meeting, Estes Park, CO, USA, 30 August–2 September 2015.
36. Keller, G.V.; Frischknecht, F.C. *Electrical Methods in Geophysical Prospecting, by G.V. Keller and F.C. Frischknecht*; Pergamon Press: Oxford, NY, USA, 1966.
37. McNeill, J.D. Electromagnetic Terrain Conductivity Measurement at Low Induction Numbers. 1980. Available online: http://www.geonics.com/pdfs/technicalnotes/tn6.pdf (accessed on 24 March 2021).
38. Rashed, M. Rolling ball algorithm as a multitask filter for terrain conductivity measurements. *J. Appl. Geophys.* **2016**, *132*, 17–24. [CrossRef]
39. Swart, P.K.; Price, R. Origin of salinity variations in Florida Bay. *Limnol. Oceanogr.* **2002**, *47*, 1234–1241. [CrossRef]
40. Stewart, M.A.; Bhatt, T.N.; Fennema, R.J.; Fitterman, D.V. *The Road to Flamingo: An Evaluation of Flow Pattern Alterations and Salinity Intrusion in the Lower Glades, Everglades National Park*; US Department of the Interior, US Geological Survey: Menlo Park, CA, USA, 2002; p. 38. [CrossRef]
41. Glibert, P.M.; Heil, C.A.; Madden, C.J.; Kelly, S.P. Dissolved organic nutrients at the interface of fresh and marine waters: Flow regime changes, biogeochemical cascades and picocyanobacterial blooms—the example of Florida Bay, USA. *Biogeochemistry* **2021**, 1–27. [CrossRef]
42. Corwin, D.L.; Rhoades, J.D. Measurement of inverted electrical conductivity profiles using electromagnetic induction. *Soil Sci. Soc. Am. J.* **1984**, *48*, 288–291. [CrossRef]
43. Lavoué, F.; van der Krak, J.; Rings, J.; André, F.; Moghadas, D.; Huisman, J.A.; Lambot, S.; Weiherrnüller, L.; Vanderborght, J.; Vereecken, H. Electromagnetic induction calibration using apparent electrical conductivity modelling based on electrical resistivity tomography. *Near Surf. Geophys.* **2010**, *8*, 553–561. [CrossRef]

MDPI
St. Alban-Anlage 66
4052 Basel
Switzerland
Tel. +41 61 683 77 34
Fax +41 61 302 89 18
www.mdpi.com

Applied Sciences Editorial Office
E-mail: applsci@mdpi.com
www.mdpi.com/journal/applsci

www.ingramcontent.com/pod-product-compliance
Lightning Source LLC
Chambersburg PA
CBHW061607210926
51850CB00009B/2242
9783036533858